Ezra Mundy Hunt

The Patients' and Physicians' Aid

How to Preserve Health

Ezra Mundy Hunt

The Patients' and Physicians' Aid
How to Preserve Health

ISBN/EAN: 9783337811853

Printed in Europe, USA, Canada, Australia, Japan

Cover: Foto ©berggeist007 / pixelio.de

More available books at **www.hansebooks.com**

THE

PATIENTS' AND PHYSICIANS' AID;

OR,

HOW TO PRESERVE HEALTH; WHAT TO DO IN SUDDEN
ATTACKS, OR UNTIL THE DOCTOR COMES; AND
HOW BEST TO PROFIT BY HIS DIREC-
TIONS WHEN GIVEN.

BY

E. M. HUNT, A. M., M. D.,

AUTHOR OF "PHYSICIAN'S COUNSELS," ETC.

NEW YORK:

C. M. SAXTON, BARKER & CO.

SAN FRANCISCO: H. H. BANCROFT & CO.

1860.

PREFACE.

The origin of this book has been the need felt in my own practice of a manual to guide as to the use of means before the physician can be procured, and as to the best methods of carrying out his directions after they have been given. In the olden time it was deemed wise to mystify and obscure the practice of the medical art, and even to claim for it supernatural aids and appliances, but a better judgment discards these occult pretensions.

There is a certain degree of information bearing upon human health and disease, which it is right should be the common stock both of physician and patient. One who tries to be his own blacksmith, carpenter, and tailor, is very apt to drive some wrong nails and make some miserable outfits, and still more, he who flatters himself that he can be his own doctor. This human frame-work is not such a simple mechanism as to require no apprenticeship, and in disease, giving medicine is not in its result such simple work as to commend itself to the trial of every sufferer.

There are then two general sources of evil, the one in endeavoring the foolish task of making everyone his own physician, and the other in shutting out all knowledge, as if "I am Sir Oracle, and when I speak, let no dog ope his mouth."

To teach how to preserve health, what is to be done in sudden cases, and what is best to be done in those milder cases, in which, whether we will or not, the people will do something for themselves, is the aim of this little work. It is a noble thing to cure disease; nobler to teach how to avoid it; and when we see how much of sickness results directly from errors which knowledge would prevent, it is natural to desire to correct some of them. It has been our effort to simplify our subjects, even at the sacrifice of flowing style and elegant diction, for, like Paul, we had rather speak five words so as to be understood, than ten thousand in an unknown tongue. A minister once gave it as a reason for preaching long sermons, that he had not time to write short ones—but we have taken time to make this a small book. By very natural expansion it might have measured more in size, but our aim has been to condense directions as much as possible, so that the means may be speedily applied, and so that advice may be quickly read and easily remembered.

We send it forth upon its mission, trusting that it will aid both patient and physician, and that its advice may be heeded by those whom it may concern. It cannot take the place of the medical attendant, but will enable you to do what you may do with propriety until he arrives, and to implicitly follow out his directions afterward. Many a child has been lost from an accident or attack which a parent might have relieved, and many a good doctor has properly prescribed a poultice, a bath, a fomentation, or injection, when, from the ignorance of the attendants, they have done more harm than good.

If this little book will enable you to do the best thing you can, when compelled to trust to yourself, and to use in a proper way the means ordered by the physician, it will aid in preserving health and restoring from disease—two objects more worthy of our attention than any save the welfare of the soul.

E. M. H.

October, 1859.

TABLE OF CONTENTS.

CONTENTS.

CONTENTS. ix

PATIENTS AND PHYSICIANS' AID.

PHYSICAL EDUCATION.

No subject, in its popular and practical bearings, is, at the present day, more deservedly attracting public attention, than that of education, and yet it is scarcely accurate to use the word without accompanying it with a definition. Meaning, literally, to educe, to draw out, to develop, as applied to man, it has, from time to time, been made to designate such modes and processes of training as the preferences of any particular age have regarded the most important elements of individual culture.

When the supply of physical wants, and the demand for warlike abilities, called for strength of muscle and firm endurance, its highest point was the development of the body. The heroes, from Hercules up and down, were all giants, and the grand triumphs of gigantic strength, whether upon the stern battle-field or amid the more peaceful sports of the Olympic games, were the themes which occupied the attention of the orator and historian, and won the admiration of all classes and conditions of society.

The same spirit formed a prominent part of Roman as well as Grecian civilization, while in turn all Europe embodied the same idea amid the wild romance of chivalry, and made of it an art, in the grand displays of the tournament.

In Greece, the academy, instead of a seven-by-nine school-room, in which little urchins were lulled to sleep or laziness by vitiated air, was a noble grove, in which old Plato taught, to healthy scholars, all that his mind could grasp of a divine philosophy. In the Augustine age, it but added to the laurels of the most learned to be able to compete amid the tests of athletic games; and, in what we call the dark ages of more modern history, noble men were made in the schools of horsemanship and sword-exercise.

One extreme is often followed by another, and too soon we find the world, in its fondness for the intellectual, neglecting the physical—in comparison, calling one mental and the other brute force. Then the sublime idea of moral education dawned with the Christian religion, and amid a just admiration of mental and moral greatness, this body, noble, and admirable, too, is unjustly left to obtain its education as best it may. Hence, amid all the boasted progress of ages, one of the most important elements of true culture has been most surprisingly neglected.

The results are anything but satisfactory. A distinguished orator, in his delineations of the burdens of society, points to disease, as one of the increasing, self-inflicted, chief; a recent book, by a female author, proves, by facts and statistics, an amount of ill health among her sex appalling to men and women both, while pale students, unhealthy artizans, complaining ministers, sick lawyers, delicate children, and other classes, making up a host of invalids, testify by battallions, that either this part of education is neglected or utterly fails of success. 'Tis true that man must die, but disease was never meant to be the rule of life, and a large share of human suffering is the direct result of parental or personal mismanagement. We need to feel that the only correct and complete idea of education is, that it is a system which develops and strengthens the physical, intellectual, and moral powers. A sound body is the befitting casket of a sound mind, and both, in healthy development, are conducive to the welfare of the soul.

Health is a blessing we often deny our children, by not habituating them to its laws, a privilege we lose for ourselves by neglect, an accomplishment we do not attain, because it requires self-denial, and a virtue which, with many, is difficult to win. Sickness is often unavoidable, but all physicians are the witnesses how much of it is the result of imprudence or ill management. While our present scope and design will not allow the discussion of a theme, in itself a volume, in briefly noticing some of the means of preserving and restoring health, we trust we shall impress the reader with the importance of the subject, and impart some information bearing upon the physical well-being and happiness of all.

EXERCISE.

It is a law of human nature, both as to mind and body, that exercise is necessary to development. In the beautiful and symmetrical arrangements by which the functions of life are elaborated, there is no provision made for inactivity. "For use," is the motto which has been inscribed in unmistakable characters, on each organ of sense and sight, on each fibre, and muscle, and atom of the physical frame-work, and he who overlooks it cannot fail to set ajar the mechanism constructed for healthful endurance.

With the very infant, action seems associated like an instinct, and tossing arms and kicking feet are not accidental and unmeaning motions, but a part of the law of growth, just as much as the beating heart and breathing lungs. The amount of running and romping which the small child will endure and enjoy, is itself an illustration of the established correspondence between nature and exercise, and yet there is no indication more frequently overlooked, by those of maturer years.

A love of ease beguiles the votaries of fashion or of wealth; a love of study or of reading too often tempts the student to neglect the casket which sustains the mind; the sedentary nature of many occupations fastens the artizan to a quiet posture, and the home employments of creation's better half, all serve as so many excuses and temptations by which the want of proper exercise becomes the undermining cause of multiplied disease. Respiration, digestion, blood, circulation, all have their appointed work, but exercise is the keystone in the noble arch of health, the indispensable support to the due performance of the functions of life. It is the first answer to the question, How to keep well? The science of physical life, how best to support and longest maintain it, is a theme deep, broad and philosophic enough for the most learned, but there are some plain truths about it which the common mind can comprehend. There must be a supply of natural want, a means of appropriating what is provided, and a method of getting rid of useless or exhausted material; and in these three specifications we have, if not the theory, at least the practical necessity of a human body. The first is supplied by industry, by the provisions of nature, and by appetite; the second by the powers of digestion and respiration, and the third by the separation of useless material from the system. With each, exercise has to do. It provides the means of sustenance, it promotes appetite and all the changes which food and air undergo, and is a still more direct assistant in ridding the system of such matter as has answered its allotted purpose and is no longer needed.

In fact, place man in a state of absolute quietude, and the other provisions of nature are not adequate to the sufficient removal of the residue left in the system, after its wants are met; this, accumulating, will interfere with all the other processes, and you must have disease resulting.

We do not propose at present, to investigate the philosophy of the necessity of Exercise, interesting and conclusive as it is, but, rather taking for granted its admitted importance, our aim will be to furnish the rules by which it should be regulated.

I.—Exercise should be out of doors.

One of its greatest benefits is, that it causes us to inhale full inspirations of air which, thus amid the million air-cells of the lungs, coming in proximity to the blood, serves to purify and send it onward circling freshly and purely through the arteries of life.

Now, if this air is impure or loaded with dust, or with the effluvia of confined locations, much of the benefit is lost. Females, especially, are apt to persuade themselves that household exercise is quite sufficient; but until they are convinced otherwise, perfect health is out of the question. It is the pure fresh air of the morning, or the balmy breeze of early evening, unobstructed and unconfined, that impart the elastic step, the ruddy cheek, and the fresh countenance of health.

Go forth amid the buds and flowers, the meadows or the parks—the delights of summer—and well clad, fear not the clear cold of winter, and thus obtain that physical strength which is secured in no other way.

II.—Exercise should not be too shy of weather.

Dull days, damp air, frosty mornings, and the fear of catching cold, are often the unlucky scape-goats of lazy indisposition or erroneous opinion. The state of the atmosphere has a bearing, it is true, upon the propriety of exposure, but it is seldom, very seldom, that colds are contracted or injury results from active exercise, even in unpleasant weather. Standing about with hands in the pocket, shivering with doing nothing, or riding on easy springs, so that scarcely a nerve can discover a motion, often are the occasion of uncomfortable feeling in unpleasant weather, but it is exceedingly rare that a person really exercising will be affected by these changes, and we believe fewer suffer from active exposure than from confinement consequent upon the fear. This very principle explains why it is that persons while actively engaged so seldom contract cold even when rain overtakes them. The system in a state of activity is less susceptible to sudden changes of temperature than when at rest.

III.—Exercise should not be taken after long fasting, nor immediately after a hearty meal.

The old advice, of a long walk or ride before breakfast we entirely ignore. Noxious vapors are never more hurtful than upon an empty stomach, after a night's repose. The system is not invigorated for exposure, and weariness soon ensues. Even the eating of a small portion of food will have, under these circumstances, a sensible effect in supporting the powers of endurance. When called out for a long ride between midnight and morning, I have always been accustomed to have at hand a small lunch, and have derived advantage from the habit.

On the other hand, after eating heartily the body needs quietude, and active exertion is not beneficial. An hour after an early breakfast, or two hours after dinner, is the best time for regular exercise, and our next inquiry must be as to the extent it should be carried.

IV.—Do not exercise to extreme weariness.

The amount that is beneficial, is very different for different persons, and is to be measured by the occurrence of fatigue. Those who return from walks or rides, so wearied that they must sit an hour to recover energy, are injured rather than invigorated. When a general perspiration has been attained, the design has been accomplished, but for the invalid, it will not do to seek even this.

With those much debilitated, it is often the case that they *need to be exercised*, rather than to exercise, and this may be done by the use of a flesh brush, and by or assisting moving the various joints of the body, as in active exertion.

No one can prescribe to another just how many miles are to be walked, or how much effort is to be made; but with these directions, each one is prepared to judge for himself.

V.—How should the exercise be taken?

This, again, must be dependent upon the physical condition of the person concerned. Sydenham, one of the fathers of English medical authority, declares that riding horseback is a sure remedy for consumption; and where the strength will permit, we even yet know of no better method of rendering active the whole framework, especially if the horse have an awkward gait. It is a mode of exercise which, without sudden exertion, calls into play almost the entire muscular system, and thus commends itself to all, both as a means of preserving and of restoring health.

Walking is also one of the best methods. Not the snail-fashionable promenade along the avenue, at the rate of a mile per hour, but the energetic, active step of a person seeking eagerly a blessing. Walk like a man of business anxious to make money—for you are after something far more valuable.

The spade, the saw and the axe are also good friends to health. With some, the gymnasium has become a favorite resort; and for those who value only what they pay for, and who need amusement or ambition to excite them to exertion, it is far better than none at all. Too frequently, injury results from the strain, and tumbling, and antics of these sports, and they are in no way superior to the simple games of children, or the natural actions of adults. The dumb-bell exercise is often beneficial to contracted chests, but even this must be used with caution, and not carried to excess.

But whatever the mode of exercise chosen, there is one universal and most important rule in respect to it— *it should be interesting;* that is, such as at the time occupies the mind. The connection between the body and mind is so curious and intimate, that even the benefit of physical exertion is modified by the mental interest felt therein. With motion and pleasure derived therefrom, the spinal cord, the sensorium and the cerebellum have to do, and these may all be performing their part when the cerebrum or brain proper, the reasoning apparatus, is comparatively at rest. By mental interest then, our reference is especially to the seat of the senses and the will, and to the lower nervous system as influenced thereby. It makes a great difference whether you walk three miles, earnestly conversing with a friend, and in no way noting your progress, or whether your attention is mainly fastened upon the special object to be attained. The sedentary student, who starts upon his ramble with his mind working, as he walks, upon his studies, is doing something that scarcely deserves the name of exercise. When the thoughts are intently busy with something else, the movement of muscles is purely mechanical, without any exerted power of the will; and it has even been contended, by some writers upon nervous diseases, with much probability, that a tendency to nervous disease is developed and fostered by these habits. Many most obstinate nervous affections consist in a loss of the power of the will over what ought to be purely voluntary motion; and as such, this moving about like mere automatons really originates such a tendency.

Boarding school misses, arrayed in pairs, talking to each other with all their might, taking the same daily, systematic, rulable walk as a task, and not as a pleasant recreation, are especially exposed to this evil. All plays creating a zeal and earnest ambition in the exercise, are most valuable, and all exerting themselves for health should be interested in and intent upon the acts in which they are engaged. The walk of the somnambulist is muscular movement, but not refreshing, and little better, action though it be, is the rotive, careless, listless, objectless exertion of those whose whole attention is at the time upon something else. Health, as the noble and most comfortable object of an earthly ambition, is so valuable that we can afford to dismiss business and even other pleasures from the mind. What the child in full earnest calls play, is true exercise; that which, while it employs the muscles and pleases the senses, demands the constant and necessary direction of the will, and manly exercise, though it may choose other forms, has here the idea and the model. With these points impressed and expanded, the adult for themselves and for those under their charge, may easily make choice of proper methods of physical exertion.

It is almost an unexceptionable rule, that the most natural are the best. Babies themselves are the best baby-jumpers, and free, natural action in man, woman and child is indispensable to the development of muscle, sinew and bone. Children, and girls especially, are sometimes too much restrained in their romping and playing, with the idea, that it is rude or not ladylike; if not, it is girl-like, and while there are, of course, times for restraint instead of primping them up on high chairs and spoiling their constitutions, let them run upon the green grass, jump the rope, chase the hoop, or ball or butterfly, and with lungs uncrowded breathe in the luscious air.

To sum up, let me entreat one and all, child or adult, the healthy and the invalid, not to be over delicate or too indolent to take daily regular out-door exercise. Remember, that far more die from confinement in houses and workshops than from exposure to air or cold, and let not a floating cloud or an habitual inactivity deter you from seeking health in the broad street, the inviting park, amid the valley's beauty, upon the river bank, or on the mountain summit, where the pure air, just come from above, is stirring in all its invigorating power.

FOOD.

What shall we eat? and What shall we drink? are two questions, the answers to which involve most important points in the preservation of health.

More persons die prematurely from abusing their stomachs than from all other causes combined. People who govern their appetites even as well as they know how to are scarce, and we should regard that man as the very personification of self-control, who never ate or drank more than he needed. Grant Thorburn, verging toward ninety years, declares he never ate enough, and although as truthful as a Scotchman, we scarcely know how to believe him. If any reliance is to be placed upon the declarations of physiologists, they show that mankind habitually eat over one-third more than their necessities require, and it must be remembered that this overplus not only does no good, but is productive of absolute harm.

It is a waste of substance, of labor and of life. It is the remark of a distinguished political economist, that more people are kept poor by their stomachs than in any other way. Did they save what they over-eat they would be better off, both as to body and estate.

I.—The great and leading point, then, that we would impress upon the reader as to food is, to have respect to its quantity.

A great deal has been said and written as to the different kinds of food, their digestibility and their adaptation to the system, and various opinions are entertained even by medical men on these points; but the quantity

of food is equally as important a subject for consideration as its quality.

This must be proportioned to the demand made upon
the system. It is impossible, by measurements of bulk
or weight, to designate precisely the quantity required
by each individual. A moderate meal for the laborious
out-door operative would be an enormous overload for
the sedentary student. Other things being equal, the
more we exercise the more we have a right to eat. The
temperature of the weather, too, has an influence upon
the calls of the system. Cold in itself consumes certain
of the elements of our food, and requires a more liberal
supply.

The growing child needs more in proportion to its
size than the full formed adult. Thus, all those dietetic
regulations which prescribe food by ounces and pounds
are erroneous, and the best regulation is a natural, well
controlled appetite—not such as has been acquired by
over-indulgence—not such as is pampered with condiments and flavorings, but such as originates from a true,
legitimate want of the physical frame-work. The rule
as to quantity should be to eat the minimum required.
When the first feeling of satiety or fullness supervenes,
it is always time to stop. For many, the appetite or
the feelings listened to at the first notice are a safe
guide; but for the habitual gormandiser seeking reform,
the only method is, never to eat quite as much as the
cravings of a morbid appetite would suggest. The
danger is in repletion, not in starvation, and he should .
aim to be on the safe side.

Whenever a change is made from an active to a

2

sedentary life, when the state of the weather is such as to prevent the usual out-door exposure and exercise, or when there is any temporary disorder of the digestive functions by a diminution of the quantity consumed, we may prevent much future trouble.

Is it not strange, how people generally, yes, even the most considerate, trifle with their stomachs? Many a one seems to prefer taking medicine to avoiding it by a proper regulation of the appetite. One cause of this is, that we do not always feel immediately the evils of excess. The corn-sheller of the careless farmer, who thrusts in the ears at random, will for a time perform as well as that of his more careful neighbor, but when the first is worn entirely out, the other will be still in working order. Though men will often tell you they can eat this or that with impunity, yet subsequent disease and declining age tell the sequel of the story. While one complains of exhausted powers and worn out apparatus when gray hairs come, the other passes down the declivity of life vigorous in the exercise of undiminished powers. You may stuff your stomach to its full year after year, and it may utter only an occasional rebuke, but as sure as effects follow causes, so sure will you reap the accumulating penalty.

There is an intemperance of eating as well as in drinking, and its results, though they may be more distant, are none the less certain. Whenever the inner man once cries enough, stop. Our own nation is proverbial for gormandizing, and over-feeding is already beginning to deteriorate the physical energies of the American people. Reform must commence with individuals, and he who learns to control his appetite achieves a victory which, with the habit once established, is no cross, and whose lasting trophies are health, happiness, and lengthened life.

II.—Food should be eaten leisurely. Rapid eating is in itself, considered independent of mastication, an evil. Mashed potatoes, apple-sauce, or anything of like character, is not so easily digested when hurled down the throat with rapidity. For this there are two important reasons. Nature has provided about the mouth certain glands, which secrete a fluid of an alkaline chemical nature which we call saliva. The food, mixed with and moistened by this, is better prepared both for swallowing and digestion.

Then, again, when at rest, the stomach is contracted, and it expands under the stimulus of food. This expansion is gradual, and if food is received too rapidly the capacity of the stomach is not prepared for it. As a part and sequence of this rule it is very important that food should be well masticated. The teeth, with their sharp edges, their beveled crowns, their nice indentations, are adapted and designed for an important purpose.

The stomach of mankind was never meant to receive pieces of meat or crude vegetables. It is not provided with an appendage like that of the cow, nor with the gizzard-like power of the fowl and the birds; nor with teeth like that of some of the lower animals.

Remember that it is meant, and only meant, to receive food after it has been thoroughly chewed and mingled with the salivary secretion. Not only is the food finely cut by the teeth, but the saliva flows more freely the more we masticate, and thus the first part of the process of digestion is here commenced, and it cannot be omitted, without imposing upon the organ a work for which it is in no way prepared. It may take more time to eat slowly and chew thoroughly, but it is time well spent. The relish of the food lasts longer, and we are not so apt to eat too much, while the material is thus reduced to such a state as to fit it for future processes.

Whatever cannot be finely reduced by the teeth is not fit to go beyond, and swallow nothing that is not thus thoroughly masticated. Any thing too tough for your teeth is much too tough for your stomach. A great many cases of indigestion and ailments which make us feel uncomfortably arise from this error. Take time to eat if you wish full time to live. ·

III.—Food should be taken at regular and not too frequent intervals. The heart is designed to beat constantly and continually until death, and, like it, the lungs, asleep or awake, at all times and under all circumstances, must be acting.

Not so with the stomach.

Rest is an essential element of its power.

It must have time to recover from the energy of digestion. This has been proven both by experience and actual observation. In the space of twenty-four hours this rest should not be less than fifteen; nor should this all come at once. Each meal should be fully reduced, the stomach thoroughly emptied, and for a time quiet, before it is summoned to renewed action.

The morning repast should be early, and may, with propriety, be the heartiest meal of the day. An old Spanish proverb says that fruit is gold in the morning, silver at noon, and lead at night, and the idea is somewhat applicable to food in general. Dinner should follow breakfast not sooner than six hours, and full as many hours after, the tea should be a light repast. It is much better to take a light supper a couple of hours before retiring, rather than partake of .the evening meal before the hearty dinner has been disposed of; for sleep interferes not so much with digestion as a stomach tired from recent exertion.

But we have said that the times of eating should be regular.

Never, in health, lose a meal, even if a slight one, and do not vary much in your time of partaking. Even horsemen understand this, and will assure you that their horses do better on regular feed, than on larger quantities at irregular times. The stomach becomes habituated to its work, and is never so ready and adapted to it as at its appointed hour; and he who eats just as convenience suits, trifles with an organ which, in old age, will sound many a painful note of rebuke.

This eating, too, between meals is one of the most ruinous habits that can be indulged. If you receive but a mouthful into your stomach, the whole process of digestion must be re-performed—if it has commenced on other food, it has to make ready for the new, or if just emptied, the mucous membrane again reddens, the gastric juice begins to flow, and it commences anew its rolling and working process; and thus, by one thing after another, is kept in a perpetual ferment. Oh! it is noble to see how nature bears up against these ills, and presses on like steam-power amid breakers; but the partaker is blinded by the very efficiency of the struggle.

The man declares it does not hurt him, for he feels not at once, always, the ill effects; but gout, rheumatism, dyspepsia, and their dire train of uncomfortable sensations, tell the story of these wrongs in advancing life.

If you are too old to reform, do, at least, teach your children lessons of self-denial and the right government of the appetite, so great a hardship for you will become a pleasure for them. It may be a little difficult at first, but when the power of resistance is once established, it gives a tone both to character and appetite, and pities the poor suicide, who stuffs and champs at all eatables whenever coming in his way.

The table is the proper place even for fruit, and in health, and ordinary activity, never eat between meals, unless it be soon after the regular repast has been taken.

Thus far we have said not a word as to what you should eat, and although most writers on this general subject dwell largely on this, we consider it the least important matter in the regulation of the digestive system. He who eats only enough, who chews thoroughly, who eats regularly at sufficiently long intervals, and takes regular exercise, may eat almost any thing eatable with impunity, and he who can not, his own experience is at least deserving of being compared with the theory of the physician, before the latter is prepared for correct advice. The old adage that, " What is one man's meat is another's poison," is not, as a general rule, strictly true, and yet there is a truth beneath this often wrongly-quoted exaggeration. The sound, healthy, well-used stomach finds no poison, usually speaking, in any of the ordinary foods; but the weak and abused organ becomes morbid in its cravings and capricious in its choice, and it is often very difficult to determine what will suit it.

The stomach of man, in a state of health, is fitted for a great variety of food, and nature, in its bounty, has provided that we should obtain and relish food derived from many sources, and prepared in various ways, and that these all should undergo such changes as prepare them for the wants of the system.

Still there is a difference, both in the mode and case of digestion, which attaches to various kinds and articles of nourishment, and it is, therefore, proper and necessary that we should place before the reader certain practical facts and principles, in respect to the kinds and digestibility of food.

I.—The first very natural division of food is into animal and vegetable. As a rule, animal food is always more readily digested than vegetable. An animal has been defined to be a digested vegetable, and it is a fact that, in its relation to man, as a food, the animal is a step in the process of digestion. The vegetables it has consumed have undergone a change similar to that necessary to fit them for human nourishment, and the meat of the animal is much more readily assimilated and appropriated than the material upon which it has fed.

There is less refuse matter from animal than vegetable food; the former remains longer in the stomach, and is more completely, as well as more readily taken up by the system. The fat of animal food is a still higher step in the process of digestion, and it is appropriated with less change than the muscular fibre. The tenderness of meat is much dependent upon fatty matter between the fibres of flesh, and the rapidity of digestion is much affected by this.

But although animal food is more digestible than vegetable, it does not follow that this is always to be preferred. Bulk, admixture, the state of the stomach, the mode of preparation, and the demand of the system for materials not contained in animal food, all point to man as a being whose supply is not to be limited by any such restriction.

II.—Another division of food, which is a natural one, is into liquids and solids.

Water is absorbed without digestion, alcoholic stimuli produce an effect upon the system without undergoing the usual assimilatory processes; but liquids such as broth, soup, and all others holding in solution materials dissolved in them, undergo a direct process of digestion, and as a rule it may be said of these that they are less quickly digested than solids. There is a separation of the solid material from the water holding it in solution which retards the commencement of the regular digestive process.

Another division of foods has reference to the chemical constituents which compose them, and which are also found in the human organism; and in this view we may speak of foods under eight varieties.

I.—Those containing nitrogen, or protein. The most prominent of these are eggs, meat, milk, and the grains of the cereals, that is, wheat, rye, buckwheat, oats, and the like. These form a most nutritious and important kind of food.

II.—Amylaceous food, such as starch, gum, and sugar. These are abundantly found in most vegetables; and also in wheat and other grains; but as the most nutritious portion of these latter is the nitrogen, they have been mentioned in the former class.

III.—Oily food derived from animals; and also from vegetables, especially their seeds.

IV.—Acid food, such as vinegar, and the various acids found in fruits.

V.—Alcoholic drinks, such as brandy, beer, wine, cider, etc.

VI.—Gelatinous food, or such as is contained in the various jellies.

VII.—Saline food, such as salt, potash, and lime, which are found in the bones and other portions of the system.

VIII.—Tea and coffee, which have a slight stimulating effect and supporting power.

Of these eight kinds, the first is the most nutritious, because it contains most of the elements of the blood; but neither of the others alone would be sufficient to sustain life in full vigor, although very important in the economy of digestion.

2*

Another division of food has reference to its particular office in supporting certain parts of the system. Thus, the nitrogenized, or protein principles, as they are called, keep up the quality of the blood, and supply the muscle.

The respiratory food keeps up the heat of the system. To this belong especially, the second, third, fourth and fifth chemical divisions of food heretofore mentioned.

The sixth variety, the gelatinous food, provides the gelatine of joints and ligaments.

Tea and coffee are believed to arrest or retard the natural waste of the body, while the saline food supports the bony frame-work.

The water, which makes up a very large proportion in bulk of all vegetables, and of the acid and alcoholic food is like pure water absorbed by the system.

Now, what becomes of the other parts after they have been received into the stomach?

The first-named compounds, the protein, or nitrogen compounds, are dissolved, and taken up by the blood-vessels of the stomach, the starch, gum, sugar, jellies and oily substances, in the form of a greyish fluid, pass into the upper intestines, and there having undergone various changes, are taken up by the blood-vessels and the small tubes, called lacteals, which are found along the course of the bowel.

Acids, alcohol, tea, coffee and saline food quickly pass into the circulation, and can be detected in remote parts of the system.

The bile, which is a secretion from the liver, performs, it is believed, an important part in the process of digesting the starchy, sugary and oily constituents of food.

In addition to these conclusions, derived from the combined investigations of many learned physicians, the opportunity has, in two or three instances, been afforded of watching the process of digestion, as performed in the stomach of the living man, directly in sight. The most interesting and instructive case of this kind occurred in our own country. A young Canadian, by name Alexis St. Martin, in the service of the United States, received a wound which perforated the stomach, and came under the charge of Dr. Beaumont, an army surgeon of high attainments. In about a year the wound healed, leaving an external opening to the stomach about two inches and a half in circumference. Dr. Beaumont felt this an excellent opportunity to perform some experiments on the subject of digestion. They were conducted at intervals, making in all a continuous period of four or five years, under a great variety of circumstances, and with ability, observation and care. It is not within the scope of our design, to trace the mode of experiment, but we subjoin herewith a table showing the mean time of digestion of different articles of food in this soldier's stomach.

ARTICLES OF DIET.	MODE OF PREPARATION.	TIME OF DIGESTION
		H. M.
Rice	Boiled....................	1 00
Eggs, whipped..................	Raw.....................	1 30
Trout, salmon, fresh	Boiled	1 30
Apples, sweet and mellow........	Raw	1 30
Venison steak	Broiled..................	1 35
Tapioca........................	Boiled	2 00
Barley.........................	"	2 00

FOOD.

ARTICLES OF DIET.	MODE OF PREPARATION.	TIME OF DIGESTION. h. m.
Milk	"	2 00
Beef's liver, fresh	Broiled	2 00
Fresh eggs	Raw	2 00
Codfish cured and dry	Boiled	2 00
Milk	Raw	2 15
Wild turkey	Roasted	2 15
Domestic turkey	"	2 30
Goose	"	2 30
Sucking pig	"	2 30
Fresh lamb	Broiled	2 30
Hash, meat and vegetables	Warmed	2 30
Beans and pod	Boiled	2 30
Parsnips	Boiled	2 30
Irish potatoes	Roasted	2 30
Chicken	Fricasseo	2 45
Custard	Baked	2 45
Salt Beef	Boiled	2 45
Sour and hard apples	Raw	2 50
Fresh oysters	"	2 55
Fresh eggs	Soft boiled	3 00
Beef, fresh, lean and rare	Roasted	3 00
Beef steak	Broiled	3 00
Pork, recently salted	Stewed	3 00
Fresh mutton	Boiled	3 00
Soup, beans	"	3 00
" chicken	"	3 00
Apple dumpling	"	3 00
Fresh oysters	Roasted	3 15
Pork steak	Broiled	3 15
Fresh mutton	Roasted	3 15
Corn bread	Baked	3 15
Carrot	Boiled	3 15
Fresh sausage	Broiled	3 20
Fresh flounder	Fried	3 30
Fresh catfish	"	3 30
Fresh oysters	Stewed	3 30
Butter	Melted	3 30
Old strong cheese	Raw	3 30
Mutton soup	Boiled	3 30
Oyster soup	"	3 30
Fresh wheat bread	Baked	3 30
Flat turnips	Boiled	3 30

ARTICLES OF DIET.	MODE OF PREPARATION.	TIME OF DIGESTION. h. m.
Irish potatoes	Boiled	3 30
Fresh eggs	Hard boiled	3 30
"	Fried	3 30
Green corn and beans	Boiled	3 45
Beets	"	3 45
Fresh lean beef	Fried	4 00
Fresh veal	Broiled	4 00
Domestic fowls	Roasted	4 00
Ducks	"	4 00
Beef soup, vegetables and bread.	Boiled	4 00
Pork, recently salted	"	4 30
Fresh veal	Fried	4 30
Cabbage with vinegar	Boiled	4 30
Pork, fat and lean	Roasted	5 30

The rapidity of digestion is so modified by various conditions of the person concerned that these proportions are not to be considered absolute rules, but serve as a guide in arriving at an approximation. From all that is at present known upon the subject of digestion, we may deduce the following views or conclusions which are valuable as guide-marks in the regulation of our appetites.

I.—Man is an omnivorous animal, that is, designed to live on both animal and vegetable food.

II.—Food should pass into the stomach in a finely divided state.

III.—The rapidity with which digestion is performed, depends upon various circumstances. Strong emotion, as anger or grief, will retard it; moderate exercise hastens it, and thus the state both of body and mind influence it.

IV.—A usual meal is generally digested in a healthy person in from three to five hours.

V.—A mixture of food is not especially objectionable, except as by variety it encourages the appetite, and often leads us to consume more than is needful.

VI.—Animal food is digested more quickly than vegetable, and solid food more speedily than soups.

VII.—Oily food is more quickly appropriated by the system than muscular fibre, when agreeing with the stomach.

VIII.—Uncooked oil is more digestible than cooked. Cream and butter are the purest of oils.

IX.—Boiled meats are most digestible, roasted next, broiled and fried the least so.

X.—Bulk is necessary to digestion. The people of cold climates who live much on fats, mix crude matters, sometimes even saw-dust with them, and thus find them more readily digested.

XI.—Milk is among the most nutritious and digestible of foods. It is considered constipating, but the chief reason is, that it is almost entirely taken up by the system, and no residue left.

XII.—With the same exertion, we need richer food in cold weather than in warm. The Esquimaux, in his cold latitude, will eat with relish his thirty pounds of whale blubber in a single day, not because he is a glutton, but the intense cold makes large demands upon the system.

.XIII.—Never eat between meals, unless extra exertion or exposure require it, and then select hearty and quickly digestible food.

XIV.—The stools from vegetable food are more copious than from meats, because vegetable is less nutritious than animal food, and more of them therefore is rejected.

XV.—As a rule, ripe fruits or vegetables are more digestible than green, and green fruit stewed more digestible than when eaten in the raw state.

XVI.—Much depends upon the articles we eat with indigestible food. Thus, a person eating nothing but cucumbers for breakfast, is much more liable to be sickened than one who at dinner has eaten an equal quantity, together with a moderate meal of potatoes, rice, etc. In other words, substances alone sometimes irritate the stomach and bowels, when if mingled with articles more digestible they would not offend.

XVII.—Smoked meats are less digestible than fresh, and of smoked or salted meats, the inner portion is more easily digested than the outer part. The inner part is preserved as much by the salpetre and the exclusion of the air, as by the salting and smoking process, and is in a state more allied to preserved fresh meat.

XVIII.—Dried fruits, as prunes, raisins, apples, etc., are unfit to eat unless well cooked, and all unbroken seeds are indigestible.

XIX.—Alcoholic stimuli, or condiments of any kind, are not necessary in healthy conditions of the stomach.

XX.—Learn from a careful experience what agrees with you and what does not, and be *the master of your appetite.*

There is a class of persons about whom special remark must be made, and these are children.

For the small child, nature has provided it food, and left us in no doubt.

God gives it not teeth, for it is not designed that it should eat any thing needing to be chewed, and its whole system points plainly to the duty of letting it alone just as nature has thus furnished it. Yet strang♥as it may seem, how many mothers there are who think the child must have more rich and hearty food. Its little stomach, no larger than an egg-cup, is supplied from the table, and the little creature at first grows fat and fast. The course would be a commendable one were you preparing it for the shambles; but if not, if its constitution be not iron, it will sow the seeds of disease to ripen perhaps only in old age, but then to ripen surely. For the first six months, mothers give the child nothing but nature's provided aliment; or if for any reason this cannot be secured, the fresh milk from a healthy, well-fed cow, diluted with twice its quantity of soft water, and slightly sweetened and warmed, will be the best substitute, because of very nearly the same composition.

Nothing besides milk is required before the seventh month, and then a little stale bread may be allowed; or the child may be fed sparingly on rice boiled in water, or mashed potatoes; but none of them should be substituted for its natural food. As a usual rule, it is best to wean children at the age of one year, before the dentition of the second year begins.

If allowed both to nurse and to eat they are apt to be overfed; are exposed to the effects both of their own errors of diet and those of the mother; and the health of both the mother and child is seriously affected by extending too long the period of lactation.

As to the frequency of eating there is a modification

in the case of children. With them food undergoes more rapid change; there is not only waste to repair, but growth to support; and the amount of exercise taken by the unrestrained child is much greater in pro portion than that taken by adults.

One piece between meals, when the breakfast is very early, may, with propriety, be allowed, but it should be of simple food, and regularity as to the time of parta king should be observed.

Other rules as to quantity, quality, and the like, apply with even more force to children than to adults, for it is now that the physical as well as moral habits of future life are being formed. Tastes like habits are acquired and strengthened by practice; and denial in early life will render self-denial easier in the future.

I would not have you, like a stoic, look with indifference upon the wants of your children; but provide them presents to please the eye rather than fire the appetite. Pastry, pound-cake, and confectionary they will take if you will furnish it, but will be equally pleased if you learn them to love that which is both good and beneficial. Thus, as far as it is in your power you confer upon them health, and enable them to choose in the years of discretion the aliments suited to them without the bias which vitiated tastes and morbid appetites make it difficult to control.

DIET.

Few words are used in a more indefinite sense than this. It is applied both as denoting the kind of food upon which a person is in the habit of living, and also as conveying the idea of a restriction in the kinds of aliment permitted in particular states of the system.

From the mass that might be said upon this subject, we shall briefly endeavor to deduce such general rules, as will prove of service to the ailing.

Our diet must bear relation to the wants of our physical natures, and in health, the appetite itself, by a natural law, craves those varieties of food most needed.

In sickness, the appetite becomes depraved, and often it requires control either in the way of restriction or encouragement. For many, the best kind of diet, in the severe attacks of disease, is to eat absolutely nothing.

All inflammations require restriction in diet at their outset, and the violence of febrile action is often effectually checked by abstemiousness. Dr. Beaumont "found that during fever little or no gastric juice was secreted, and, consequently, food only served to irritate the stomach and the whole system."

But it is hard to convince people that they can live even a short length of time upon nothing, and many an attack of disease might be warded off, or its severity abated, by a strict avoidance of food at its very outset.

In the future course of disease there is, at one period or another, one of three indications.

1. Either to reduce the system. 2. To support it gradually and mildly, and with those kinds of food most easily digested; or, 3. To overcome a tendency to sinking or debility by the most supporting and sustaining aliments.

Where we desire to reduce and thus to abate the violence of disease, we have already said that nothing is

often the best thing you can give; but it is often desirable for the physician and the friends of the patient to
know, in case something is desired what is next to
nothing. Arrow-root or corn starch, prepared with
water, is very slightly nutrient. Apple-water, rice-
water, orange-juice, and pure cold water may serve to
quench the thirst and satisfy the desire, without affording much nutrition. This class of sick persons usually
desire very little food; and hence the variety need not
be great. To such, gruel is often offered as light diet;
but any thing that will fatten pigs we do not regard as
low diet.

A small piece of toast, well buttered, is another of
these so-called light foods; but if you will analyze wheat
flour and butter, you will find them to contain the most
hearty ingredients of food.

If there is a real necessity for abstemiousness, do not
deceive yourself with the idea you are practising it
when using oil or meat, fibre or flour, or milk or eggs,
in any form.

The second indication, viz.: To support the system
by kinds of food which are easily digested, and yet not
over-stimulating, is a very common one, and worthy of
careful notice.

Jellies, plainly made, either from animal matter or
fruits; fresh, ripe, mellow fruit; boiled rice; stale
bread crumbled in milk, if rich, reduced with water;
eggs boiled soft, by being broken in hot water; boiled
fresh fish; the soup from oysters or clams, or from boiling the fibre about the bones and joints, this being rich
in gelatine; are especially indicated. Among vege-

tables, the tomato, the carrot, cauliflower, asparagus, and roast potatoes, well cooked and plainly dressed, occupy the first rank. Among meats, boiled mutton, tender roast beef, squirrel, birds, and the soft part of fat oysters, either raw or slightly cooked, are both palatable and supporting, but, where there is much excitability about the system, and a tendency to continuous fever, these are seldom allowable.

Food to this class of sick should be given at regular intervals, and not too frequently.

III.—Where there is great waste of the system, and a tendency to death from exhaustion, we find the great demand for stimulants, and the most nutritious diet.

Alcohol and the essential oils, as found in brandy or other stimuli, are often here required, and cream and milk will assist much in the support.

In order that the effect of a stimulus may be permanent, it is always best to combine it with a nutrient, and to give it in small and frequently-repeated doses, so that the sinking reaction, which is a part of all quick stimulation, is overcome by repeated quantities, until a tone and strength is secured, which will sustain life.

The best of meat, butter, and stale-bread are here admissible; and, after all, good food is the greatest of tonics and the best of stimulants, and the object of other remedies is to prepare the system for its reception, digestion, and appropriation.

DRINKS.

Thirst, as well as hunger, is one of nature's calls for supply to the wants of the system, and the right choice of our drinks is no less a matter of importance than eating, and first in our estimation comes the pure cold water. The pump and the cow, says Hawthorne, are the two greatest replenishers of thirsty humanity. Nothing so grateful to the lips parched with fever as a draught just from the spring or the deep, cold well, and alike in sickness and in health, it is the best beverage of life.

Unless rendered too cold by ice, or taken in too large quantities, or drawn through lead pipes, it seldom will do harm. Two or three suggestions, however, are proper in respect to it.

It is not best to drink much at meal time. The gastric juice itself, and the liquid in the food, furnish sufficient dilution to the solid matter, and where the drink is very cold, it interferes directly with the process of digestion.

This cannot go on with the temperature of the stomach lower than 100° Fahrenheit; and it is quite easily reduced below this. Drinking, like eating, is to a great extent, a habit, and many use far more than the real wants of the system require.

Regularity in this, as well as in eating, is desirable, unless extra demands are made by salt food or perspiration. Usually the best time for drinking is about two hours after the meal, when the food has mainly undergone its stomach-digestion. Drinks slightly warm agree best with weak stomachs, and the abstemious are sometimes injured by their cold-water breakfasts.

Milk diluted either with cold or warm water, is well suited to those who need something more nutrient as a drink than water, and who yet cannot bear the more stimulating compounds. In respect to coffee, the almost universal morning drink, many and varied views are entertained as to its effect. It certainly has slight tonic properties, and unlike the alcoholic stimuli, retards the usual waste of the system. To those who are sedentary in their habits, who are nervous or inclined to disordered stomachs and headaches, it is undoubtedly very injurious, and should not be indulged in at all. All children are much better off either upon milk or water. By none should it be imbibed in large quantities, for excess in it, as in other things, will tell sooner or later upon the system; but, with plenty of milk and well sweetened, we are disposed to regard it, for persons of active habits by far the least objectionable stimulant that can be employed. Proof is not wanting, that in its use persons will endure extra labor and fatigue better than with alcohol, and the permanent effect is not so deleterious.

As to tea, we feel that more serious objections can be urged against it.

From its mode of preparation it is liable to injurious adulterations; the mode of drying it often imparts to it even chemical qualities, and the tea-leaf just from the tree, is quite a different thing from the metallic infusion associated with it.

As a medicine, it is good for those not accustomed to

its use in case of severe headache of a nervous character, but in those habituated to it, it often produces the
restlessness which in the uninitiated it will allay. We
hope that the time will come when the leaf of some
harmless domestic tree or shrub will be substituted in
its place, and we thus at once be provided with a
cheaper and more wholesome article.

Cocoa and chocolate are so nearly allied that they
may be noticed together. They are chiefly valuable
because they impart a pleasant flavor to the milk they
contain. Both of them, like tea, are most extensively
adulterated. From one of the best reputed preparations in the market, we have, when cold, skimmed
enough mutton-grease to turn the stomach of any invalid. The prepared shell is the only form in which
the cocoa should be used, and this forms an excellent
substitute for tea or coffee, because prepared with milk.

As to alcoholic stimuli, a book would be required to
discuss their merits and demerits. It is the rule and
law of their action that they stimulate but to depress.
There is always a reaction. If the pulse rises under
their influence, it afterward sinks below the standard.
Their ill effects, habitually used, may not be perceivable
at the time, but they leave a mark on the constitution
which shows itself in the fading years of age.

Not only as philanthropists, but as having regard
merely for your physical welfare, we would say never
as a habit use the intoxicating beverage in any form.

Alcohol has its place as a medicine. If you have
overloaded your stomach like a glutton, or are suffering
from indigestion, a little brandy may help you through

the trouble, but it is not because it is good for you always. You do not reason so about other things. Opium may help you in a colic, but on this account you do not take it every day. Except as a medicine, alcohol is never needed, and for three reasons it is a medicine you should not prescribe for yourself.

First, it is often adulterated, and medicine being bad enough at any rate, you should make choice of that most pure.

Second, it is a dangerous remedy. More have died from it than from strychnine, or hydrocyanic acid. The habit acquired has ruined thousands whom all mankind would regard as stronger-minded than you or I.

Third, you can in all cases where you ought to prescribe for yourself take something in its place. Ginger, cayenne pepper, or mustard tea, will usually answer as well.

In ale and porter, tonic properties prevail over the stimulant, and these are often valuable tonics. As to other drinks, you can form an idea of their value by considering them as made of water and that which flavors them. Thus, the various acid drinks are good in fever, because the water is cooling, and the acid aids to allay thirst. Emollient teas, such as flaxseed, soothe by the mucilage they contain, and thus each combines the separate effects of the water and the ingredients used.

What shall I wear? is often the question of fashion; should it not be also the question of health? There are two extremes; to avoid dressing too much, and not enough. If the dress be too warm, it causes an unnatural heat of skin, engenders a dryness unfavorable to free circulation, or a perspiration which renders us more liable to colds.

If too scanty, it exposes us to sudden changes which, from the sympathy existing between the skin and internal organs, is apt to be felt in vital parts, and thus affect the centres of life.

Consumption is, with us, the most potent sceptre of the destroying angel, and the chest and throat should therefore be well protected.

For the former, the open vest is too exposed in winter; and for the latter, a wash with cold water each morning, and a warm silk cravat are the best protection. Large outside neck-bands of woollen, while they keep out cold, often produce so much heat as to cause us to leave the neck moist, and liable to cold when they are removed.

The value of flannel, mostly or entirely wool, for children and adults, as the inner garment of the whole body, cannot be over-estimated. It acts as an equalizer of the temperature and its ready absorption of the perspiration guards against sudden colds. The fireman and engineer wear it, and this fact, with their experience, answers in full the common objection that it is too warm for warm weather. After once used, few complain from this cause, and in our changeable climate, there is far

3

more danger in leaving it off than from inconvenience
in using it.

A new or thinner flannel may, with propriety, be
used in the summer, but it should not be entirely cast
aside.

The feet should always be kept comfortable by such
materials, as while they preserve warmth, will not keep
them wet with perspiration. People differ much in this
respect, and the preference for woollen or cotton stock-
ings, for thick boots, or India rubber shoes, should be
tested and governed by this consideration. Wetness or
dampness of the feet from any cause, or prolonged cold,
will often surprisingly affect the state of the whole body.
A proof of it is that many a headache is relieved in a
few minutes by soaking them in hot water.

With the body protected by an inner covering of
flannel, and the feet kept comfortable; so far as health
is concerned, it matters not what the quality of the
outer garments may be, so long as not so many as to
overheat, or so few as to chill.

As to the covering of the head, the hair is its natural
protection. The bare-head Dutch rarely contract colds,
and the open hats of our American ladies, although
objectionable, seem to show at least that these are not
very injurious.

We believe, upon the other extreme, that men err as
to the head-covering much more than females. Caps
should never be worn without a small opening in the
top, to permit evaporation, and the hot air to escape,
and even the tall hat should always have a ventilator
in the top. If you doubt, suspend a small thermometer

in the crown, and, after wearing it a few hours, see to what a temperature it is kept.

No wonder that colds in the head are contracted by such close confinement; and I am confident of having relieved many an ache by this suggestion. By this little expedient the hat will be equally comfortable, and the temperature more equable.

As to tight lacing, everybody that thinks knows that the lungs do not need squeezing or the bowels cording, and that it is not sensible for man or woman to make ridges between them, and so we shall not spend words upon this point. Girls should wear their dresses suffi-ciently high in the neck to allow broad straps for the under-clothes resting high on the shoulder; for if not, the constant hitching which their sliding off causes, often gives rise to a habit which results in a drooping of one shoulder and a partial curvature of the spine. Children are more easily chilled than adults, and gener-ate heat less rapidly, and should, therefore, in propor-tion, be more warmly clad, with flannel as the inner garment.

Make up your mind in your own dress, and that of those you control, to follow the fashions only when they are not prejudicial to health; but, when they tell you to ride, on a cold December's night, in short sleeves, a thin, low-necked dress, no hat, and as good as no shoes, or to dress your little ones in clothes only reaching a little below the waist and a little above it, be sensible enough to dress neatly, without seeking the heights or depths of fashionable perfection. Though it may differ from the ideas of some, we kindly suggest that, it is better to have the name of being sensible than that of wearing the latest pattern.

SLEEP.

"Night is the time for rest.
How sweet, when labors close,
To gather round an aching breast
The curtain of repose—
Stretch the tired limbs and lay the head
Upon our own delightful bed."

Do not take too much of it. It is impossible to say just how much; but indulged in to a great extent, it enervates rather than refreshes.

Seven hours will suffice for nearly all, and if in doubt, try this, until you are convinced that more is needed. Cease labor or logical reading half an hour before bedtime. Go to bed early and get up early, is a rule upon which I cannot improve. The morning sun ought never to rise before you. Severe labor, mental or physical, protracted late at night, wears more on the constitution than when commenced at reasonable time in the morning. The one is as a hearty supper, the other as an early breakfast to the mind.

Sleep not too warm, as it interferes with exhalation from the skin. A mattress, or straw bed, is far healthier than a mass of feathers, and a pillow entirely of feathers is quite objectionable.

Always have a night suit, so as to wear nothing to

bed that you have had on during the day. Thus your
garments are aired, and cleanliness greatly promoted.
If any exception is made to this rule, it would be in
favor of the doctor, and he will not ask it. Never lie
long in the morning after waking. If so, you are apt to
doze and dream, to the injury of mind and body. In
summer, a short nap after dinner is often refreshing and
beneficial, and, at any time, better than a late rising in
the morning. These suggestions observed, together with
temperance, exercise, and regularity, will usually make
the rising as sweet as the repose, and aid in conferring
and preserving ruddy health; so that, in your own
experience, you may find "balmy sleep" to be "tired
Nature's sweet restorer."

VENTILATION.

Air is the most indispensable support of life, and its
purity has a vital bearing upon health. It is through
its direct agency that the blood undergoes its greatest
changes. We hear much about medicines that purify
the blood, but after all good fresh air as received into
the lungs is its greatest purifier. There are two kinds
of blood—venous and arterial; or, as you may call
them, impure and pure blood; the one circulating in
the veins, the other in the arteries. Arterial blood is
made impure in travelling through the system because
its oxygen is imparted to the body, while impurities in

the form of carbonic acid, and other substances no longer needed, are imparted to it, and by means of smaller vessels, or capillaries, as they are called, it passes into the veins, and back again it comes to the lungs. Pure air, and pure air alone, has the power to take away this carbonic acid, and substitute in its place the oxygen which *it* contains. If this process is entirely interfered with, death usually takes place in four minutes, and carbonic acid is the poison which causes it.

If you will confine a crowd of people in a room to which pure air cannot get access, you soon obtain just the state of things which occurs when a charcoal furnace is lighted in a tight apartment. In either case the air becomes charged with carbonic acid, a light will burn dim for the same reason it will in a deep well full of foul air, and to those confined within death is inevitable if they do not seek relief.

Under such circumstances hundreds and thousands have perished, poisoned by their own breath. In the "black hole" of Calcutta one hundred and twenty Englishmen died from this cause in a single night, and in 1798 a passenger ship came within the Isle of Wight, the captain having for safety during the wind storm ordered all below deck, and tightly fastened and sealed the hatches. When these were removed every passenger was found dead. So much of death lurks in the confined breath of every mortal. These are extreme cases, and although such results can only occur when fresh air is completely excluded, they teach us important lessons. The oppression felt in crowded or close rooms, the flickering of the lights, and the feeling of drowsiness often

ensuing, are no small indications as to the evil effects of
this foul air. But unfortunately we cannot trust always
the notice of our senses. The person habitually exposed to a vitiated atmosphere, after a time becomes to
some extent insensible thereto, but his system does not
become insensible to the ill effects. The student may
work in the dissecting-room until he notices not the odor
of putrefaction, but this does not make it healthy.

The investigations of scientific men, sustained by
practical experience, have established the following important facts:

I.—Although carbonic acid is not the only impurity
in air already breathed, it is, nevertheless, the chief
source of evil, and we may measure the ill effects of
foul air by the excess thereof.

II.—When air contains more than one-half per cent.
of carbonic acid, it is injurious to breathe it for any
length of time; increasingly so up to six per cent., and
beyond that, if long continued, will prove fatal.

III.—Air once breathed is not fit to be reinhaled
until freed from impurities by free admixture with the
great ocean of air which surrounds us in the form of
atmosphere.

IV.—The amount of air inhaled varies with age,
health, exercise, &c.; but the average amount changed
at each inspiration, is about fifteen cubic inches, and
for an adult there are about eighteen inspirations to the
minute.

V.—None contend that less than four feet of pure
air a minute is sufficient for the aeration of the blood
as an average for men, while experiments in connection

with the British House of Commons, and other careful trials and calculations, seem to denote ten cubic feet as the smallest amount per minute accordant with perfect health.

Taking our fourth and fifth specifications as the basis, over six cubic feet would be the amount designated, and this is a small allowance for an active man.

A room ten feet long, ten feet broad, and ten high, contains $10 \times 10 \times 10 = 1,000$ cubic feet, and ten persons, in less than seventeen minutes, $10 \times 6 \times 17 = 1,020$, will have once breathed all the air in the room, and this, by the amount of carbonic acid it already contains, has been thus rendered unfit for further respiration.

This is making no allowance for other unhealthy excretions passing off with the breath or from the skin, all of which go to make it more injurious still.

Now, with such facts in view, sit down and calculate for yourself the evils of deficient ventilation, and with close rooms, rendered still more objectionable by over-heated, as well as vitiated air, wonder not if colds and ill health supervene, and if from your seven by nine sleeping room you issue in the morning unrefreshed, and the three children in the trundle-bed dream, and roll, and cough croupy, wonder not.

It may be said, in answer to this view, that no room is so tight as not to admit some fresh air through the doors and crevices, and this is very fortunately true, but in cold weather especially, these are all kept carefully closed, and they do not secure a sufficient ventilation.

The practical question, then, comes home to us, How

is proper ventilation to be most completely and economically secured?

1st. Rooms can easily be ventilated when they are not being occupied.

The sleeping-room, in which we spend nearly one-third of our lives, should surely never be the sitting-room. In summer and winter, each day, the outside air should be admitted thereto, even if it be but for a few minutes, for dampness or coldness is better than the foul air of the night's repose retained, and the airing may be given in the morning, when the room is en tirely unoccupied.

The sleeping room should always be one of the best apartments in the house, and either so large that the pure supply of air will not be all used during the night, or by open windows, doors or ventilating apparatus, the escape of the carbon and the entrance of the nitrogen and oxygen of which pure air is a mixture, should be secured.

2d. The room occupied by the family during the day, can also be easily ventilated early by an open window; but as it is frequently inconvenient to aerate it by open doors during its occupancy, other means must be devised.

Windows should be made to drop from the top, and should extend nearly or quite to the ceiling. Impure air, though in itself heavier than pure, when warmed ascends, and a ready exit for it is not afforded, unless the point of ventilation is as high as the air can reach.

Chimneys form a very natural and important means of ventilation, and the old open fire-place, with its bla-

3*

zing fire, bore no unimportant relation to the health of
those it warmed.

As the difficulty of proper ventilation is chiefly experienced in cold weather, and as the best methods of accomplishing it are by availing ourselves of certain laws
and effects of heat, we must glance briefly at their
relationship.

Heat, in itself considered, has little to do with the
purity of air. You may sleep or live in a cold room,
and yet the air become impure, or you may have it
warm, and yet ventilated so as to be pure and healthy.
Extremes, either of heat or cold, ought to be avoided,
and an even temperature of 70° Fahrenheit is usually
sufficient for comfort and health. A very dry heat is
unfavorable to proper exhalation from the skin, and a
basin of water *warm enough to evaporate,* should be kept
upon the hot stove. The gas of coal, or even of burning wood retained or thrown out into the room, is very
injurious, and means should be devised to prevent this,
by a good draught. All these points are worthy of attention, because, through neglect not only is the carbonic
acid of the breath retained, but more is furnished from
these defects, and thus a double accumulation secured.

But allowing that these external sources of foul air
are duly attended to, and that no decaying animal or
vegetable matter about the premises is generating this
poison, how shall we be rid of that constant and unavoidable supply which is sent out from the lungs of
every breathing creature?

One, or all of three things is essential.

I.—There must be a way of escape for the impure air.

II.—There must be a mode of entrance for fresh and pure air; or,

III.—The two must be combined, so that there shall be a stream of pure air flowing in, and of impure flowing out.

Even the first plan alone, though imperfect, will accomplish much. The impure air passing out, leaves a vacuum to be supplied, and pure air will find its way in more easily than if the former is retained.

The second method alone, will expel some of the impure air, but will not so readily displace it, as it will rush into a vacuum, and therefore is not equal to the first method.

Perfect ventilation requires the combination of both methods in such a way that there may be a free going out of impure, and coming in of pure air, and the great difficulty is, to accomplish this without causing a draught, which will be uncomfortable, and unfriendly to health.

Theoretically this can be avoided, by having a great many very small apertures through which fresh air can enter, and foul air escape, but practically, the thing is not perfectly easy to accomplish, and at the same time to keep up a regular, and comfortable degree of heat.

Our limits will not permit us to pursue this important subject in detail, and we shall therefore condense what we have to say in a few important suggestions.

Chimneys with an opening directly into the room furnish an important means for the egress or getting out of impure air, but to render them perfect for this service, three things are necessary.

I.—There must be a good draught. In order for this, the heat in the chimney must exceed that of the room. In every chimney, before heating, there rests a volume of the external air which prevents the warm air of the room from escaping, so that if either, it serves rather as a means of letting in cold air, instead of removing that in the apartment. In fact, when there is no motive power, it does very little toward ventilating the room, unless the outside air is warmer than that of the room.

But when fire is made, the heat causes a vacuum, so that the currents of air from the room flow toward any opening, whether by stove, fire-place, or an aperture made in it, and thus it becomes a most effective means of ventilation.

A stove is far less effective than a fire-place for the pipe often does not much more than allow room for the gas and smoke from the fire, and if it be air-tight, the value of this mode of ventilation is entirely lost.

II.—The opening in the chimney for ventilation will be more effectual, if it be near the ceiling. Although foul air at the same temperature with pure air falls, yet as it comes heated from the lungs, it rises, and diffuses itself most in the upper part of the room. The point of ventilation should therefore be at least higher than the height of persons occupying the room. If lower, since foul air diffuses itself among that of the room, much will escape, but not so perfectly as with a higher opening. A cheap and ready-made method, therefore, of ventilating your room, is to leave a brick or two out of the chimney near the ceiling, so that thus there is an opening from the room. This may be at the side or in

front, and a picture-frame may so be hung over it, as
not to interfere with the flow of air. If you do not
choose to have this, or an open fire-place, let the pipe
of your stove enter above the mantel, and instead of an
r —————— c elbow, have the stove-pipe run into
another at right angles to it, thus,
so that the end c enters the chimney,
and the end r is open into the room.
The diameter of this pipe must be
greater than that of the pipe s, so that
there may be opportunity for the escape of air.

s

To this arrangement, and the former one of a hole in
the chimney, it may be objected, that sometimes the
state of the air will be such as to cause soot, smoke,
etc., to be blown out, through these apertures, into the
room.

In a properly-constructed chimney this can scarcely
occur, for the draught will prevent this, or, if not, there
are two contrivances that will. The one is, to have a
fixture known as a cowl, such as that invented by Mott
of New York, fastened on the top of the chimney, which
suits itself to the direction of the wind.

Another more usual plan is, to have a slide over the
opening in the chimney, or pipe, which can be opened
or shut at pleasure, or better still, one which will open
and shut itself as occasion requires. The ventilating
valve of Dr. Arnott acts in this way. "It is, in principle, a small weigh-beam or steel-yard, carrying on one
arm a metallic flap to close the opening, and on the
other a weight to balance the flap. The weight may be

screwed on its arm, at such a distance from the axis or centre of motion, that it shall exactly counterpoise the flap. Then, if the draught is stronger downward than upward, it will close the valve and prevent smoke; if stronger in the upward direction, it will act with its free power as a ventilator. Although the valve, therefore, be heavy and durable, a breath of air suffices to move it, which, if from the room, opens it, if from the chimney, closes it, and when no such force interferes, it settles in its closed position." Thus, in a warm chimney, the air from the room can go out, but any backdraught is prevented. In the general use of the valve, the weight should be so screwed on, as to leave the valve raised, except in case of a strong downward draught, which will more than restore the equilibrium.

A third plan is, to have the air flue alongside of the chimney flue, and thus separated from, but heated by it. This may commence from the ventilating aperture, and extend to the top of the chimney, and in this case a simple slide, if any, will answer for the opening.

A metallic pipe in the centre of the chimney, with its base resting upon, and connected with, the opening near the ceiling, would be better still, as being thinner, and surrounded on all sides by the flue of the chimney, the air within would be better heated, and thus a more perfect vacuum be secured.

Under ordinary circumstances, any of these simple contrivances are adequate to the removal of foul air, and they are so cheap and easily applied, that at least every sitting room daily occupied, should be thus provided. If so, fresh air will usually find its way in, through crevices and keyholes, sufficient for most houses.

Where, as in public buildings, there are a great many persons together in a close room, there may need to be, also, provision for supplying pure air; and this must needs be done without creating a heavy draught, or without chilling the room. To avoid a draught, air must enter at several small points, instead of a single large one; and in order to prevent it chilling the room, it must be warmed before entering. Hosking and Reid of England, and Curtis, Bull and others, of our own country, have suggested and put in practice important principles for the accomplishment of these objects, which may be understood by a reference to their explanations and contrivances, or by consulting scientific architects.

Where there are rooms in which there are no chimneys, these may be connected therewith by tubes, and thus a mode for the escape of foul air be provided, and where houses are heated by furnaces, instead of fireplaces, an arrangement like Bull's furnace will secure both heat and ventilation.

We believe the public are far from appreciating how much ill health is occasioned by imperfect ventilation in sitting rooms, sleeping rooms, school-houses, churches, and public assembly rooms, and a careful consideration and expansion of the before-mentioned principles will, as we hope, draw attention to the subject.

This luscious air so plenteous, free and pure, Oh! let it not be banished from our lungs, when by such simple means its benefits may be secured.

The skin, as the external covering of the whole body, is a network well worthy of our most careful analysis and care. Made up of cells the most complete, with thousands of bloodvessels running in every direction through it, with multitudes of nerves looping up into it so thickly that you cannot thrust the point of a needle between them, and with its sweat and oil-glands, it is among the most wonderful parts of the human organism.

Upon it heat and cold, vapor and smoke, acids and emollients, things beneficial and injurious, act with a power which is transmitted to the very seats and centres of life. The number of pores in the skin of a man of ordinary size is seven millions, and the number of inches of perspiratory tubing alone would be equivalent to a length of twenty-eight miles. The body gives off through the skin by exhalation in the form of insensible perspiration, etc., every twenty-four hours, an average of a pound and a half by weight per day. It is, so to speak, an outspread surface of bloodvessels, and nerves, and glands, and tubes, and pores, in itself an outer world of wonderment. Is it surprising, then, that its condition is intimately related to that of the whole system? Cool it too suddenly, and the lungs are oppressed by a cold; wound it, and the heart fails in its action, and faintness supervenes; inflame it, and the stomach is all in disorder; destroy a portion of it, and the nervous system sinks at the shock.

In fever, its fountains of moisture are dried up, and

in almost every ailment it shows its interest by a corresponding change. So far as we can, how important that we should keep it in proper condition, so that it shall not become diseased ; so that its circulation shall not be impeded ; so that its sweat-glands shall be unchecked, its oily secretions not allowed to remain until rancid ; and its whole surface be aided in performing its functions aright. If you can succeed in regulating the temperature of the skin, and in keeping all its parts in working order, it is the experience of all medical men that disease is the more easily subdued. One of the important methods of preserving it in its proper state is by *ablution*.

We all seem to recognize the propriety of washing the face each day, and yet, with the exception that it is a little more exposed to dust, it stands in no more need of care than other portions of the body. Covered as the latter is by clothes, it is sometimes the case that the escape of some of the excretions is even more interfered with. Washing is necessary to the perfect health of the whole body. In no other way can its minute pores be kept open and free, the grease dissolved, and the whole surface enabled to perform its functions. This ablution, too, has, also, a reflex action. It affects the body internal as well as external.

It keeps open the canals by which substances no longer needed in the system, are disposed of, and thus each organ is enabled to perform its part, unclogged by the noxious or useless materials it has at hand. It is surprising how many persons neglect this plain and sensible duty. Why, I know some so-called respectable people, whose whole bodies water has not touched in a single year. If in summer they get into a pond or visit the river it is a fortunate circumstance. Such a state of things is positively awful, and no wonder that fevers and a dire course of medicines are necessary to free the system of its pent up humors.

Do now take a bowl of water and wash yourself from head to foot, or if it be cold, warm your water to suit you, get along side of the stove, and wash away, and resolve never to let more than a week or two pass without a thorough cleaning. It will soon become a luxury, and you will rejoice in the sweet feeling of cleanliness. See to it, that the children pass through the same process, and thus establish for yourselves and them a stepping-stone to health.

BATHING.

Under this topic we shall not dwell upon the virtue of cleanliness, for this is secured by ablution, but water has other offices and effects beside that of a mere cleanser. Used in various forms, it has a direct agency in the promotion and preservation of health.

It is not, in all cases, to be regarded as a harmless remedy. It is not like the bread-pills of the doctor's wife, " which, if they didn't do any good, didn't do any harm." By pouring a stream of cold water upon a child's head but for a few minutes, you can kill it; by remaining too long in a cold bath, you can occasion internal inflammation, and give the system a shock from which weeks will not recover it ; by using the hot bath too long or too frequently, you may cause congestion and nervous prostration, which will result in immediate death or long protracted ailment. So, after all, this bathing is not a very simple matter, for you deal with an agent which, according as it is judiciously or injudiciously employed, may prove one of the most potent of tonics, or the most powerful of depressents.

We shall speak of four kinds of Bath.

I.—The Cold, or that in which the thermometer (Fahrenheit's) does not rise above 60°, and it is very rare that a degree of cold below 40° is desirable.

II.—The Tepid Bath, between 62° and 96°.

III.—The Warm Bath, between 96° and 98°. Ninety-eight degrees is the usual warmth of the body, as found by a thermometer placed in the groin or arm-pit.

IV.—The Hot Bath, from 98° to 106°.

Of all these there is one modification—the shower bath.

Now the effects of water, and the rules for its use, are vastly different, according to these different temperatures, and the way in which it is applied.

The Cold Bath is one of the most popular, and there is one rule universal, in respect to it, that should ever be remembered. Unless there is reaction, it always does harm. If you come out of it cold and shivering, and after drying, rubbing and dressing, remain so for any length of time, it has surely done you harm. The shock has been too severe ; it has acted as an astringent upon the skin, rather than opened its pores. Whenever, therefore, you take a cold bath, see to it that you get a speedy reaction. The skin should be rubbed briskly with a rough towel. Both the rubbing and the exercise will do good; or if not able to bear the exercise, have an assistant. It is not every constitution that will bear the cold bath, and we believe many injure themselves by its too frequent and indiscriminate use. Begin with the water a little warm; do not stay in too long, and go on to pure cold, if you find the system bearing it well, and invigorated by it. It is a tonic, the value of which is variable, and to be tested by your own careful experience.

The Tepid Bath is in general better suited to invalids, and to those whose constitutions are not very firm. It, too, is colder than the skin, but the difference in degree of cold causes less shock to the system, and reaction is more readily established. Here, too, there is need that the body quickly assume its usual temperature, and hence, when there is much debility or want of activity in the circulation, even the tepid bath is not admissible.

It is, however, more generally applicable than any other kind of bath; but remember, when I say tepid, I do not mean hot, or cold, nor what your hand would bear; for a child or lady's hand tests heat quite differently from that of a farmer or mechanic. I mean a temperature between 62° and 96° by your common thermometers.

The Warm Bath is of a character quite different from either of the other two. While they are prominently tonics, this is more of a stimulant, and afterward a depressent. Where you wish to relieve some internal

part, to soften the skin when dry and husky, and cause
more blood to circulate upon the surface, the warm bath
is of great value. It is, therefore, of much service in
internal pains, or congestions from any cause. In a
severe cold, for instance, it tends to promote free action
from the surface, and to relieve the pressure, by attracting blood to another part. The reaction to be guarded
against in a warm bath, is a feeling of depression and
weakness. A little of this aids in accomplishing the
object, but, if carried too far, may produce much prostration. Whenever, therefore, there is paleness, or a
sense of faintness ensuing, the person, instead of being
rubbed with flesh-brushes, must be wrapped quietly in
a blanket, and placed in a comfortable bed.

Where the symptoms of disease are not severe, a partial warm bath, such as of the feet or arms, or a hip
bath, will answer every purpose. While the early part
of the day is the best time for a cold or tepid bath, the
evening is better for a warm one.

The Hot Bath is only applicable in cases of disease
where an agent of considerable power is required.
When you wish to produce revulsion, as when there is
severe pressure upon some internal organ, a quick hot
bath may relieve it very soon; but there are two possible evils to be guarded against, the one is, not to continue it so long as that the heat is transmitted to internal
organs, and the other, to see to it that there is not too
much consequent depression.

The hot bath, in general, needs to be a very short
one, three or four minutes often sufficing to obtain its
benefit, and many have seriously suffered from continu-

ing them too long. As the cases requiring the hot bath generally soon come under the physician's care, it is not needful that I should say more about them. Have, then, definite ideas as to the degrees of temperature, denoted by the terms in general use, and of the designs to be accomplished, and you will find pure water not only the noblest of drinks and the best of washes, but a most valuable and effective remedial agent.

As to the shower bath, cold, tepid, warm, or hot, its effects are of the same character as each of these already mentioned, but it has a much-augmented power.

I exaggerate not when I term it a tremendous agency for good or evil. A stream or streams of water, pouring from a height upon the head, and over the body, will often revive from a sudden stupor, when nothing else will. The snoring drunkard will respond to the working of the pump-handle, when nothing beside will attract his attention. In a sudden flow of blood after confinement, cold water poured thus over the abdomen, for a moment, is often a proper and effectual remedy. Warm water poured over a part where a sudden pain has seized, as in lumbago or pain of the hip-joint, often speedily relieves, and thus we might by many instances illustrate its power.

The Vapor Bath requires such care, and is so seldom needed, that we shall not here describe it, as it should never be used, except under the careful direction of the attending physician.

We have thus summed up all we have to say about bathing, except to assure you that water is not the only good thing, and while you should respect those fully alive to its value, and recognize it as one of the best of remedies, do not drown yourself in hydropathy, with its one-idea name, or allow yourself to be enveloped in wet sheeting, when you can get a reaction in a much more simple and less objectionable way.

COLDS AND COUGHS.

"Nothing but a cold," is just what more people die of than of any other one thing, and a cough is something that should never be trifled with. Cough may arise from other causes than a cold. A slight mechanical ir-ritation of the throat may produce it, a disease of the liver or heart may give rise to a slight hack, so that all coughs do not denote a difficulty of the lungs: but a cold, and in general a cough is connected with an irri-tation of the lungs. People talk of feeling sore over the stomach, instead of the lungs, and swallow all sorts of remedies, as if indeed the gullet and the windpipe were one and the same.

What is a cold? It is the checking of perspiration, and the answer conveys with it many lessons, both as how to avoid or cure it. From the body, we have al-ready noticed, there is constantly a perspiration, a giving off of useless material, sometimes perceptible, but more frequently unnoticeable. Now a draught of air, or sudden change of temperature, acting upon an exposed surface when in a state of perspiration, is very likely to produce a cold. This may effect a particular part, or the whole body, but the most common result is an irritation of the lungs.

If, then, you are warm or sweaty, never sit down in the draught. Do not keep your coat on when hard at work, and throw it off when done, but pursue precisely the opposite course. If you wear flannel you will not be so liable to contract cold, for this absorbs the moisture from the skin, and prevents the cold air from acting directly upon it. When very hot, always cool off gradually, and do not expose yourself to a much lower temperature. We are more likely to get a cold at evening than at other times, because then the temperature of the air sinks rapidly. We catch cold more readily with empty stomachs than after feeding, more quickly asleep than awake. It is a sudden change from heat to cold, or cold to heat, that usually causes the trouble. If you get wet, so long as you remain actively at work, so that the evaporation does not reduce the heat of your system, there is no danger of catching cold; but if you cease work before dry, always change every wet garment, and rub the skin with a dry towel.

Cold is often occasioned by sitting close to the stove in a very heated room, until the skin is parched and dry with heat, and then going out into the cold air. If you have a cold, how shall you get rid of it? Remember it is the checking of perspiration, and if you can establish this again, and keep it established, you recover from your cold. The first step of a cold is made upon the delicate net-work of the skin. When the first symptoms supervene, go into a comfortably warm room. Either take a warm bath or soak the feet, drink a bowl of hot bone-set, catnip or flax-seed tea, put a mustard plaster over the chest, wrap a warm blanket about you, and go to bed.

Do not get along side of a red hot coal-stove, and parch your skin, as persons when shivering are very apt to do ; but use these better means of getting warm without drying the surface, already suffering for a want of action.   If in the morning the cold is not relieved, a very mild saline cathartic, as Seidlitz powder or salts will be of service if the bowels are constipated, and a teaspoonful of syrup of ipecac every two hours, or as much of the powder as you can hold on a three-cent piece, dissolved in water, and taken each time like syrup, will be of service.

When a cough is accompanied with a severe sensation of tightness over the chest, or with pain in the breast or side, and especially if a chill supervenes, you have reason to fear inflammation of the lungs, and medical aid should be promptly procured.

Always make it your business to take care of a cold until you get clear of it, and do not trust too long to your own judgment.

The following is a valuable, pleasant and safe cough-syrup in cases of sudden colds.

Take Syrup of Tolu,
     Syrup of Ipecac,
     Comp. syrup of squills, } Equal parts.
     Paregoric,
     Sweet oil.

Mix.   Dose : a teaspoonful every three hours.

4

A good Domestic Cough Syrup which you may prepare yourselves, is made as follows :

Take   Horehound,

            Catnep,

            Tansy,

            Hyssop,

            Hops,

of each a half a teacupful, boil in two quarts of water
down to one quart, and strain ; then add of

            Comfrey,

            Elecampane,

            Gum Arabic,

            Liquorice root,

            Sarsaparilla,

            Balm of Gilead buds,

each half an ounce; of sugar a pound and a half, and
molasses a pint and a half; boil down again to a quart,
then strain, and add one gill of best brandy. Dose,
a tablespoonful, three times a day.

Regularity of life in every respect, is especially desirable for those inclined to lung affections; and often,
where the usual so-called cough remedies do not avail,
good living, exercise, bathing, pale ale, and a combination of iron with one of the vegetable bitters, may often
be prescribed for you with advantage. Insidious as
consumption is in its formative stages, it is often more
under the control of the patient than most diseases, and
one with the panacea of self-control, and a physician
who, though he may occasionally give medicine, feels
the greater importance of prescribing the habits and
regimen of his patient, need not despair.

SUDDEN BLEEDING.

This may occur without previous notice, and prove fatal before a physician could be obtained, and there-fore all should know something of the means to which they may resort to control it.

When raised through the mouth, it is usually either from the lungs or stomach, and, so far as your imme-diate treatment is concerned, it matters but little which. If amounting to much, provide at once a hot foot bath, spread a plaster of pure mustard and place it over the whole chest, and administer a teaspoonful of salt, either dry or in a little water. The patient should be placed in a cool room on a mattress or straw bed, and warmth applied to the extremities. Keep the room quiet, have no one in it except the necessary attendants, keep the person as quiet and as much in the recumbent posture as possible, with the head low, and endeavoring to avoid coughing or straining. If the hemorrhage is severe, feed him pieces of ice. These directions attended to, will probably control it until you can obtain further aid. In a first attack, the patient himself is apt to be much excited, which increases the amount; but he should be assured, as is the case, that the immediate danger is not so great as it seems.

For hemorrhage from the uterus, pouring cold water over the abdomen, and bandaging *tightly*, are the most ready and effective means.

Alarming bleeding from deep wounds may result either from the cutting of a great number of smaller vessels, or that of a larger one, such as a main vein or

artery. For wounds of the smaller vessels, the application of cold to the part until the bleeding partially
ceases, then bringing the sides carefully together, and
doing them up, will usually suffice to control the
hemorrhage. Thus pressure is made, or the clots of
blood collecting fill up the wound. When a large artery is wounded, there are two expedients to which you
may resort; either making pressure against the bone
above the wound, or tying the artery at the point of
cut. If, for instance, the leg or arm is wounded, a handkerchief tied tightly about the leg near the groin, or about
the arm near the shoulder, and a small stick placed in
them and turned so as to press very tightly, will usually
answer until more intelligent assistance is procured; or
if you find the open mouth of a vessel spurting out
blood, if you can get a strong thread about it you will
do no harm.

Bleeding from the nose is not uncommon, and is difficult, sometimes, to control. In moderation, it is often a
benefit, and should not be too quickly restrained. When
you wish to stop it, just place the ball of the thumb
against the nostril from which it comes, and hold it
there, so as not to allow the exit of blood. Keep
drawing the air up forcibly through the other nostril,
and apply cold to the back of the neck, or over the
bridge of the nose. The blood will soon coagulate and
stop the bleeding vessel; and, if let alone, in a day or
two the point from which the blood emanated will have
healed. If this does not check it, dust, or ashes, or
pieces of cobweb, or pulverized alum, or a little tannin,
snuffed up the nose, will assist the blood in its coagula-

tion.˙ These may be used, either alone, or upon a piece of cotton thrust up the nose and allowed to remain for a time. Bathing the feet in hot water, and perfect quietude, are also of advantage.

Persons subject to nose bleeding should wash the head frequently with cold water, and by warm clothing and well protected feet, see to it that the extremities are kept comfortable.

LEECHING AND CUPPING.

These are simple and often effectual means of abstracting blood, or relieving internal congestion ; and as the physician frequently cannot remain to oversee them, the method should be understood by other attendants. In leeching, the animal used is what is known as the blood leech, a little creature of foreign birth, imported expressly for this purpose. It has three small teeth, by which it makes a triangular incision into the skin, and sucks out the blood therefrom. The common leeches of our ponds, which are caught sometimes by boys wading in the water, will often, after they have been kept a little while in clear water, answer a good purpose, but in general the Swedish leech is preferred.

In order to apply a leech, the part must first be carefully cleansed with a little warm water. Then various methods may be resorted to, to enable it to perform its work. Two or three may be placed in a wine-glass or small tumbler tipped over the part, so as to confine

them to a particular surface; or a single one may be
taken hold of by a small linen rag wrapped about its
middle, while the head, which is the smaller end, is
directed to the desired point. Touching a little sugar
and water to the part, or thrusting the leech suddenly
in a little weak 'vinegar and water, or beer grounds,
may induce it to lay hold. Placing a little blood on
the surface, will often avail when no other plan will.
After the leech has taken hold, it should be left alone,
and suffered to fill itself to the full, when it will drop
off. If it clings too long, sprinkle a little salt upon the
surface of its body, but never pull it off, as by this
means the wound it has made is sometimes rendered
troublesome, and even a part of the long tooth of the
leech left in the opening. The leech may then be made
to disgorge the blood, by taking firmly hold of its tail,
and stripping it in the direction of the mouth, or salt
may be sprinkled upon it, which will cause it to reject
its meal. Then rinse the salt off from it, and place it in
cool water. Leeches can be kept best in a glass filled with
rain water, covered over with a thin muslin, and placed
in a cool, dark place; or, as by druggists, in moistened
black earth.

After the leech is removed, the wound of the patient
requires some attention. If, as is usual, it is desired to
continue the bleeding, the parts should be bathed with
warm water, or a warm poultice placed over it. In the
case of children, the after-bleeding needs to be watched,
as it may be so great as to occasion faintness, or great
prostration. When it is desired to check the flow, bathing with cold water, pressure for a moment with a bit

of ice, thick gum-arabic water, a little lint, or powdered alum, or chalk, or tannin, or any of the articles directed to be used in bleeding from the nose, will usually avail. If not, the doctor must be sent for, and he, by the use of caustic, or a small needle, will probably succeed in controlling it.

Cupping is, in some respects, superior to leeching. It costs less, draws the blood just from the point you desire, is more of a counter-irritant, and you know just how much blood you obtain. On the other hand, to some parts it cannot be applied so readily as a leech, and, on a thin skin, leaves a scar sometimes objectionable.

To perform it, you need a cupping glass, a scarificator, and a torch or spirit lamp. With the ingenious, an inverted wine glass, a pen-knife, and a burning piece of paper will answer as well. The design is to exhaust the air from the glass by the heat, and to apply it quickly to the part. By suction, the skin is raised, and more blood remains at that point. If, then, the glass is removed, the part wounded, and the exhausted glass reapplied, the blood will flow out. The scarificator generally used consists of several very small knives, which, by the action of a spring, all wound the part at once. The spirit torch is used, because it emits no smoke, and the cupping-glass may be of almost any shape or size.

Choose the place to which you wish to apply the glass, which must needs be nearly a flat surface. Wash the skin with a little warm water, so as to render it soft and pliable. Place your cupping-glass on the spot, light your torch (which may be made of a little cotton, placed on the end of a stick, dipped in proof spirit, and lighted), raise the glass, thrust this quickly under, thus exhaust the air, and reapply the glass. Thus, the skin will raise under the suction. This is known as dry cupping, and is of value, and resorted to even where no further process is attempted.

When, as usual, it is desired, besides this counter-irritant effect, also, to abstract blood, you remove the glass, scarify the part, and reapply it, as directed above. Several trials and some failures may be necessary to make you dexterous, but the tact is easily acquired. The most common error is, that in applying the cupping-glass, after it has been exhausted of the air by the torch, it is thrust upon the surface, so as to create a draught, thus again filling the vacuum upon which your success depends. It should be applied quickly, yet gently, the glass being already held as near the part as possible, and, as it were, slid on, as the torch is withdrawn.

After sufficient blood has been obtained, the glass, by a slight pressure of the skin on one side of it, is easily removed, care being taken to retain the blood within it, or to catch it by means of cloths. A little oil may be smeared over the wounded part, or, if it is desired to continue the bleeding, fomentations will be of service. In foreign practice, leeching and cupping are performed by professed adepts, and even the common striped barber's pole is the relic of olden times, when the barber was also the bleeder. Were the simple operations of leeching and cupping more generally understood, so as not to trespass upon the time of the physician, we believe they would be much more frequently prescribed for relieving internal congestions.

Other methods of abstracting blood are only safe under the immediate eye of the surgeon.

BURNS.

These are so common, so painful, and often so sudden and serious as to demand the most careful and immediate attention.

There are three degrees of them.

1st. Where the skin is merely reddened.

2d. Where there is vesication, that is, where the skin is loosened or raised, as in a blister.

3d. Where more than this, the parts beneath are deadened or charred.

Burns from steam are generally the worst. It quickly affects a large surface, and is often inhaled into the lungs, as in explosions, and the person destroyed in this way. If exposed to steam, throw a woolen blanket over you, and you may rush through it in safety. If you have to pass through a blazing fire, first puff gently out all the air you can from the lungs, take in a full, quiet inspiration of fresh air, and rush through, not inhaling any, and not breathing your supply faster than is necessary. Woolen will protect you better than cotton or linen.

I.—To a slight burn, in which the skin is not broken,

4*

you can apply nothing better than a little weak alcohol or sugar-of-lead water, or a mixture of a teaspoonful of linseed oil, and one of lime-water, which it is well to keep on hand prepared. Laudanum and sweet oil, equal parts, mixed, will control the pain, if severe.

II.—When the skin is broken or raised, as in a blister, the symptoms are more marked and the treatment more important. Do not break the skin at all, or any more than it is, as this is the best dressing, and prevents the air affecting the tender parts beneath. If the person has been scalded in a part covered by clothing, dash tepid water on it before you attempt to remove the dress, and then do it very carefully, so as not to tear the raised skin. There is often considerable prostration from burns, especially in children, and the general symptoms need attention almost as soon as the local. If there is coldness or faintness, you need not fear to administer brandy and water, equal parts, every few minutes, by the teaspoonful, until reaction comes on. Warmth to the feet is also desirable. There is nothing better you can do for the burn itself than to smear it by means of a feather, with equal parts of turpentine and sweet oil, or the lime-water mixture we have before spoken of. To be prepared properly, the lime should be dissolved in water stirred, and then allowed to settle. Then drain off the water, mix it with an equal part of the oil, and beat it as you would eggs to make cake. Protection from the air is very important for a burn, and the value of oily substances depends much upon this.

Another important matter is, to reduce the temperature of the part gradually, and hence such an article as turpentine, containing an oil with a slight stimulant, is very valuable as a dressing.

Cotton wadding or cotton flannel may be placed over the dressing, so as to reduce the temperature by degrees, but not directly next to the burn.

Where there is much uneasiness, pain, or restlessness, after a few hours, it is always safe to give paregoric to a child or laudanum to an adult. The doses we shall notice under that head.

Another good mode of treating burns is with what is known as the water-dressing. This consists in wetting a linen cloth with tepid water, placing it over the burn, and covering it with a piece of oiled silk. Whenever the linen becomes dry by the evaporation, it should be again wet and reapplied. Even where this is used, a slight previous application of melted lard, alcohol, or turpentine is desirable.

Burns should not be made to heal too rapidly; first, because you are almost sure to get proud flesh in them if they do, and, secondly, the skin contracts, and leaves a worse scar. This is one of the disadvantages of many of the patent pain-killers and stimulants, and a recommendation to the water-dressing.

In a bad burn, always place yourself under the care of a good physician, as they need careful attention and management.

III.—Where the true skin is destroyed, as is seen by the whitened or charred appearance of the burn, matter is always, sooner or later, formed, and you have more constitutional disturbance, and more drain on the system, from the suppuration which must necessarily ensue.

At first, stimulants, as warm turpentine and the like, may be applied. The tepid water-dressing should be an early application, and this may give place to a flax-seed or slippery-elm poultice. As soon as the dead matter has separated, the water-dressing will be again of service, and the future treatment will be the same as that hereafter described as applicable to healthy ulcers.

Sometimes we have burns from acids, or from alkalies, as lime, potash, and the like. They need just the same subsequent treatment as other varieties, but, by knowing three things, or two, just at the time, you may check their extent and severity.

If burnt by an acid, apply quickly an alkali, as soda, or chalk, or saleratus, lime, or magnesia, dissolved in water, and strong enough to be tasted.

If burned by potash, or lime, or other alkali, apply vinegar and water, or other mild acid, and then you will need to be treated as for other burns. Lime getting into the eye often causes great pain and trouble, but the speedy use of the mild vinegar wash, and afterward bathing the part with hop tea, will prevent or check the inflammation.

Burning Fluid, as the mixture of alcohol and spirits of turpentine is generally called, is in such common use for lighting purposes, and so frequently the occasion of accidents, as to be worthy of special notice.

On account of the intensely inflammable character of the mixture, the flame runs rapidly over a large surface, and burns from it are therefore apt not only to be severe but extensive. Prevention is proverbially better than cure, and surely much better than cases like these, and for this, two cautions are especially to be enjoined.

1. Never use a glass fluid lamp. They are liable to be easily broken, and over nine-tenths of the accidents occur from these.

2. Never allow a fluid or camphene lamp to be filled when it is necessary to use a light in order to do it. Have at hand a few candles, or some other light, so that if there has been neglect, some other mode of lighting may be temporarily used.

It is the general opinion that a fluid lamp will explode, and that accidents from them occur from this cause. Such is not the case. An eminent chemical professor of New York city informed me that he investigated a large number of these accidents, by noticing every case of explosion reported in the papers, and visiting the house, and in no case did he find that a real explosion had taken place. My own experience fully confirms this testimony. Sometimes, by a crevice in the glass or metal, the parts may be separated, or more frequently the heat of the flame melts a part of the substance, so that the liquid pours rapidly out; but it is in no sense an explosion, and a knowledge of this fact should lead to much greater presence of mind than is generally manifested. If a lamp suddenly flame up, do not shake it about, or drop it at your feet, or toss it suddenly in the air, so as to spill its burning contents all over you, but at once set it quietly down, and keep clear of the flame. It will not explode itself upon you, and will get upon you in no other way save by the force of gravity. Even if a little has been spilled upon you, no flame is more easily blown out if the amount of fluid is not great; and if it has caught to the clothing, do not run around and create a breeze to fan the flame, but wrap yourself, or be wrapped, at once in a woolen blanket, or carpet, or shawl, or any thing of the kind nearest at hand. Thus let effort be made to smother the flame, and let water be dashed over the burning part. The after treatment is the same as that already directed for severe burns. With even common precaution, how-ever, accidents will rarely occur, for they are as much dependent upon carelessness, fright, and thoughtless-ness, as upon any inherent wickedness in the fluid.

BRUISES.

Under this term we designate those cases in which some hard substance coming in contact with the flesh, produces a disarrangement more or less severe of the soft parts. This varies from the production of a little blue-ness to a complete mashing and killing of the part. The first application to a bruise should be cold water, as this keeps the part from swelling, and drives the blood away rather than permitting it to settle. But this must be superseded by warmth and moisture as soon as there is any considerable pain. The warm-water dressing, or poultices, frequently changed are the best means of pro-viding this. These make the skin pliable, restore animation to parts almost deadened by the shock, and thus aid in their restoration.

Where there has been much blood thrown out, as is known by the blackness, leeching will have to be resorted to, to relieve the over-laden vessels. Never keep a bruised part in motion. In whatever portion of the body it may be, it needs rest, and imprudent working or walking may easily convert a simple bruise into a most troublesome swelling. After the active symptoms have subsided, there still may remain some enlargement, and if in the vicinity of a joint, some stiffness. Under these circumstances, friction, as rubbing with the hands, or a coarse flannel, is often very beneficial, and in tedious cases, in addition to this, you may use some stimulating, and oily liniment, such as you will find noticed in the chapter on liniments.

These plain and simple methods will usually be effectual in restoring the parts; but if the quantity of effused blood is so great, that the bruise breaks, in this case, the poultice has not been amiss, and the sore left after the discharge will readily heal.

If the bruise is beneath an unyielding part, as the nails, or the thickened skin of the hands or feet, the pain is much greater, on account of the inward pressure on the nerves, and warm applications need to be most freely used so wet as to soften the part. If it is under a nail, it is well to scrape this with a bit of glass, so that it will yield to the pressure, or even to make, if the pain be intense, a small aperture through which the effusion beneath may escape.

Where a bruise occupies a place where the skin is

loose and flabby as around the eye, it will sometimes enlarge enormously, and the man who has received but a slight blow at this point will still parade before the public a large black eye. Now, black eyes are often admired, but not just this kind, and I advise you to keep clear of them if you can, but if not, use cold water upon them at first, apply a leech, if very painful, and comfort yourself with the fact that it is not so serious a matter as it looks to be.

When a part is bruised which naturally hangs downward, as the hand, it should always be supported in a sling, so that the blood may not settle so much in it. It is never best to open a bruise unless matter forms beneath it, and this you may know by a throbbing pain continuing for several days, and a softening and pointing of the swelling, which may then be opened; or by continued poulticing be made to open itself. Even a common stone bruise by opening, and allowing the blood to discharge and the air to enter, is sometimes converted into a troublesome sore. It is far better to apply heat and moisture, and, after pain ceases, a slight stimulant to cause the blood to be absorbed. As a secondary result of bruises, we often have boils and felons; but we shall have occasion to speak of these hereafter.

SPRAINS.

While bruises refer especially to soft parts, sprains have reference to ligaments and bones.

You have probably noticed what is called the whit-leather in an animal's neck, as it is frequently seen in the meat upon the table. Now ligaments are represented by this. They are the cords by which muscles are attached to bones, and the tough ropy texture by which the bones are connected with each other. Some severity is always needed to sprain or overstretch the parts, and when hurt they are slow to forgive the injury. Hence sprains have always been considered serious injuries, and, when severe, recovery is more tedious than in fractures. They always occur at joints, and hence they have to do with a very nice and susceptible part of our framework. The substance affected in a sprain is hard to hurt, and hard to recover.

The first point in the treatment of a sprained joint is, to keep it at perfect rest in the easiest position, which will be that which relaxes all the muscles about it. The first application should be water, of a temperature as cold as the patient can bear; but if the pain increase very much in severity, this must give place to warm dressings, poultices, and soothing appliances, such as laudanum, or hop-tea mixed with the poultice.

Often it is necessary to abstract blood, not only by means of leeches, but also by bleeding from the arm. If the pain is excessive, do not hesitate long in sending for your physician, for soon serious damage may accrue to the joint. Among warm applications, the leaves of the wormwood steeped in hot vinegar and applied to the sprained part, often seem both grateful and curative. For the first few days, the diet should be low, and the bowels kept free by seidlitz powders or salts.

After the constant pain is relieved, there will often remain stiffness and slight pain on motion, but moderate use should be persevered in, and now is the time to expect advantage from rubbing and stimulating liniments. One teaspoonful of tincture of Spanish flies, two tablespoonfuls' of sweet oil, and two of 95 per cent. alcohol, furnishes a ready and valuable mixture thus to use.

Pouring cold water over the joint until it occasions slight pain, will also, at this period, help to strengthen it. If it remains swollen, it should be supported by a flannel roller, and by careful use it will in due time recover its wonted power. A reference to the article on poultices will inform you as to the best methods of preparing them when required.

FROST-BITE.

The nature of a frost-bite is so well known to all, that we need not spend much time in describing it. The fingers and toes, the ears and the nose are the parts most liable to be affected. Cold reduces the heat and circulation of the part, until at length, these, together with the nervous impression, cease, and the temperature sinks below the freezing point. Persons are often partially frozen before they are aware of it, and in very cold mornings caution is necessary, or ere we fear it, the white mark of the frost-bite rims the edge of the ear, or sits supreme upon the tip of the nose. We not unfrequently hear of instances of persons frozen stiff, and without consciousness of their danger, because without pain, dying in that condition.

A person exposed to extreme cold should always keep in motion. It has always been the testimony of those who have just escaped being frozen to death, that rather a pleasant sensation comes over them after the first shock of cold, and is soon followed by a desire for repose.

The account given in Captain Cook's Voyages, of Dr. Solander, Sir Joseph Banks and three companions, forcibly illustrates this.

Exploring the hills and swamps of Terra Del Fuego, they were overtaken by piercing blasts of wind, accompanied by snow, and found it impossible to reach the ship that night. Dr. Solander, familiar with the laws of exposure, called his companions together and counselled them to keep active. "Whoever sits down," said he, "will sleep, and whoever sleeps will wake no more." Yet Solander himself was the first who felt this inclination, and persisted in lying down, and notwithstanding the earnest efforts of Sir Joseph Banks, he went to sleep, with another beside him. He had rested but five minutes, when word arrived that a fire had been kindled at a little distance, and by the utmost persuasion and exertion he was urged on to it, but his fellow-sleeper "waked no more."

The great rule to guide you in the treatment of frozen

persons or limbs, is just this. The change of the part
from a frozen to a natural state, must be gradual. If
you kept the part at an extreme of cold equal to that it
has already suffered, it would of course remain frozen.
If, on the other hand, you bring it to the warm fire, the
change of temperature is so sudden as to destroy the
part. If a person has been frozen so as to be insensible,
bring him into a cool room, remove the clothing, and
rub the whole body with snow or cool spirits. If
warmth and sensibility begin to return, your business is
to moderate, rather than to hasten it. Rub the patient
dry, place him on a straw bed with little covering, and
if reaction comes on quite fast, let a window be raised.
Remember, after the thawing process has commenced,
there is little fear but that it will be continued, but
more, that it will be too much hastened. A teaspoonful
of brandy, or an injection of warm water now and then,
will at this stage be found an effectual aid. Where only
a part of the body, as one of the extremities, has been
frozen, the same general rules of treatment apply. The
limb may be rubbed with cool appliances, in a room of
very moderate temperature, and thus its heat gradually
and slowly increased. I have seen, as do all physicians,
many a case where a hand or a foot has been lost, by
warming it by the fire while frozen stiff. It should at
first be exposed to a temperature but a few degrees
above freezing point, and thus gradually made warmer.
Afterward, the limb may swell and tingle, and blisters
rise on the surface; but there will be no blackness or
sloughing. It will remain some time weak and trouble-

some, but after a few days the use of the following liniment will hasten recovery :

Alcohol, 2 tablespoonfuls,
Spirits of camphor, 1 tablespoonful,
Sweet oil, 2 tablespoonfuls,
Laudanum, 1 teaspoonful.

Mix and apply each day.

CHILBLAINS.

Chilblains are of something the same nature as frostbites, but less severe. They result from sudden changes of temperature. We have seen that a part actually frozen, and brought suddenly to the fire, mortifies. A part chilled, and brought suddenly to heat, is weakened by the sudden change, the surface is unnaturally irritated, and you have itching, burning, and discomfort generally. Persons then say they have slightly frosted the part; but the truth is they have too suddenly warmed it, when only very cold. Pimples, and itching of the face, and uneasiness of the hands or feet, often arise in this way. The child after playing in the snow runs in, and places its hands close to the stove; or the man with cold feet, instead of working them about, or walking around the room, sticks them against the stove, burns his boots, and in a few days is pretty sure to have troublesome feet, and the best kinds of itching ears are obtained somewhat after the same fashion. Not only then should frozen bodies not be brought to the fire, but when very cold, while we may seek a warmer temperature, we should not toast ourselves at the stove. The after-effects of rough skin, and uneasy sensations, etc., though often not recognized as resulting from this, nevertheless do. Washing the face or hands in hot water, when very cold, or in cold water when very hot, often irritates the skin, on the same principle.

But perhaps you already are fitted out with a chilblain, and if so, you need the pound of cure rather than the ounce of prevention. If it has not broken, wear a flannel or woolen stocking, do not sweat the feet by tight boots or rubbers, and bathe the part each night with a liniment made of equal parts of spirits of turpentine, spirits of camphor, and laudanum.

If the chilblain is much inflamed, with a throbbing pain, or is discharging, you cannot do better than to keep dressing it with warm poultices of linseed meal, to which a teaspoonful of the above liniment may be added, and the after-treatment is that for an ulcer needing stimulation, for which see remarks on ulcers. The boot should be made to sit loosely over the part, and the chilblain be frequently oiled, until recovery is fully established.

CHAPPED HANDS.

These, like chilblains, arise from sudden changes from heat to cold, or cold to heat. When moist or wet, the action of cold is more severe upon the skin. If you wish to get the hands or face chapped, warm them hot by a stove, and dash them into cold water, or wash them in ice water, and hold them quickly to a very hot fire. These sudden transitions of temperature, especially in connection with moisture, cause the skin to break, and give rise to crevices troublesome to heal.

But, if you have already been imprudent, or, from a natural tenderness of skin, have furnished yourself with rough, uncouth hands, how shall they be cured?

If merely made harsh, without being so as to bleed, bathe them with sweet oil each night, and wash them with soft water and castile soap in the morning, and dry them with a soft towel, without rubbing. The common potash soaps are very bad for such cases, and should never be used. Where the crevices are deep, the part should be soaked, and the hardened skin on the edges carefully scraped off, and then the same course pursued, the hands being protected from exposure by silk gloves. Where the grooves are so deep as to bleed, the edges should be touched with lunar caustic, and, if disposed to spread, be brought together by adhesive plaster. A little red precipitate ointment rubbed on the part will often facilitate recovery. When the lips are sore from a similar cause, people will eat and talk as usual, and hence it is difficult to keep them still long enough to unite the edges of the crevices. If you can accomplish this, and pursue the general course above indicated, they will soon be restored.

CORNS AND BUNIONS.

These are so common, that to many this is a feeling subject. They are always produced either by pressure or friction; or, in other words, they always result either from a too close confinement, or a rubbing of the part. A corn is formed just in this way: A boot or some hard substance pressing upon the part, causes a thickening of the outer skin. This becomes rough and hard, and inflames the sensitive skin beneath. The pressure continuing, one layer after another is formed, the upper part being broader than that below. Thus, so to speak, a kind of peg of skin is produced, which from its hardness and corn-like shape, keeps pressing deeper and deeper, thrusting aside and irritating the parts below, and thus producing great uneasiness.

A tight boot, or a very loose one with narrow toes, will occasion them. In the first, the pressure is constant. In the second, the foot, in walking, continually presses forward, and thus bruises the tops and sides of the toes against a hard surface.

Boots or shoes, therefore, should not be so tight as to keep up a constant pressure, or so loose as to allow the foot to slide much forward. Have them to fit smoothly and snugly over the instep, and thus you avoid the forward pressure, and with sufficient width over the toes,

"That doleful crop of pedal maize,
Called, by the vulgar, corns,
Will never flourish there."

But if already flourishing, what shall be done to get

rid of them? Soak the part in warm water, and cut off all the hard outer skin until you just begin to feel it, but not enough to draw blood. You now perceive a little centre, or, as the people call it, the seed or core. Dig this out as deep as you can without pain or blood, by means of a sharp-pointed knife. Drop in this a single drop of tar, or instead, and sometimes better, get of your druggist a little piece of sal-ammoniac, and powder enough to fill the hole thus made. Then take a small piece of flannel or soft kid, cut a very small hole in it to correspond with this point, and fasten it on by means of a narrow strip of adhesive plaster passing round the toe. This should be done *every day* until the corn is destroyed. You thus protect it from pressure, and provide something to destroy the corn. The use of tar or sal-ammoniac powder, will often be sufficient without the covering, and a cure will often be effected. In addition to this, you must of course cease wearing feet coverings that do not fit, and thus get cured not only of the corn, but of pinching your feet out of reason, or wearing boots so loose over the instep as to jam your toes at every step. If a corn becomes much inflamed, it must be washed in warm water and poulticed until soreness subsides, before you can use any other means for its eradication.

In Bunions, the surface is more extensive, and the irritated skin does not thicken before nature makes her complaint. The part swells, becomes red, inflamed and painful, and thus forces the patient to attend to it. They are most frequent over the instep or upon the joint or ball of the large toe. A moderate degree of equable

5

pressure does not give rise to them. When upon the instep, it is usually because the boot is tighter over one portion, as over one side of it, than the other. A proper measure and fitting will avoid all this. A bunion once fairly produced, even after the soreness has subsided, is apt to leave an enlargement of the foot at the particular part, and thus mars its symmetry and beauty.

If a bunion begins to make its appearance, the first thing to be done is to relieve it from pressure, and keep it relieved. This will often at once check its progress; but if not, a warm linseed-meal poultice will be serviceable. After the part has been relieved, to hasten its restoration, and render it less sensitive to pressure, it is well to bathe it with weak brandy and sugar, or, if there is any danger of this getting within, a weak tea of tannin or white oak bark will serve in its place. It should also be frequently oiled, to prevent friction, and the use of the liniment suggested for chilblains will often be found of advantage.

WARTS.

These are growths upon the skin, either attached to it by a kind of neck, or having a broad base, and more deeply set therein. They sometimes seem to come and go without any known cause; and their cure is often attributed to beads, and weeds, and chains, and trinkets, the stuff that fancy loves to dote upon.

Where the wart is elongated, a silk string drawn tightly

about its base will usually, with safety and certainty, remove it. This should be tightened daily until it is separated. Where all, or any part, cannot be removed in this way, on account of the broadness of the base, it should be soaked in warm water, and all the loose surface skin removed, as much as can be done without bleeding. Then let it be rubbed with the powdered sal ammoniac, spoken of for corns, *daily*, and this persevered in will usually cause it to disappear. If not, severer caustics, as nitrate of silver, or oil of vitriol, may be used.

Where the warts are large, the best method is not to use severe caustics too freely; but, if their removal is determined upon, to resort at once to the knife.

Where they are situated upon exposed parts, as the face, care must be taken lest the remaining scar be unsightly. In applying sal ammoniac, or any caustic, the wart should be soaked in warm water, and the loose or discolored skin gently removed before reapplying the remedy. Success depends upon following up with the remedy daily.

CURVED AND TROUBLESOME NAILS.

It is not unfrequently that a patient presents himself to his physician limping and looking sorrowful, about nothing but a toe-nail curved too much into the flesh, but any one who has ever suffered with one will think it no small matter. The nail thus acts precisely as a splinter would tied fast at the same point, and each step and motion serves to increase the irritation.

Generally the cause is an error of art and not of nature. Boots too small at the toes press the. flesh at the sides of the toes until it overlaps the nail. Frequent and close cutting of the nails, too, favors their yielding to any pressure, and thus the nail is gradually made to penetrate the margins of the toe. This gives rise to a secretion, which, becoming dried and hard, increases the difficulty, and often suppuration and the formation of proud flesh are the consequence. As in a great many other cases of minor surgery, the prevention is much simpler than the cure, and parents should teach their children to guard against the difficulty.

When the feet are bathed care should be taken to cleanse the nail, where it comes with its edges in contact with the flesh, yet not to irritate it, and the trimming should not be very close in the corners. If disposed too much to curve, the edges may be raised with a bodkin, and a bit of lint or woolen tucked under the corner from time to time will be of advantage. If the difficulty increases, scrape the nail with a piece of glass in its middle portion, so that the part will give to pressure, and become less rounding.

By these precautions, resorted to in time, and shoes wide at the toes and snug over the instep, you will avoid all trouble. But if this is already fully at hand, a different course must be resorted to. The toe should be well soaked in warm water, and if there is proud flesh, this may be removed by a little burnt alum. As soon as the intense pain is subdued, a small piece of wool, or strip of linen, should be inserted under and around the edge of the nail, and kept in place by a piece of adhesive plaster. The nail should be allowed to grow out long, and especially the corners should not be trimmed; for though relieving a little for the time, it prevents you from curing the difficulty. By allowing the linen tucked under to extend over between these and the side flesh, irritation will be prevented. This should be attended to daily. The upper surface of the nail should be scraped as thin as possible without pain. If, at its point of greatest curvature, instead of scraping, you take a round file and make a groove, so that the arch will yield, you may sometimes effect a cure.

One of the great difficulties in curing this evil is, that the patient is unwilling to keep quiet, and hence the part is irritated by frequent motion. Rest will much facilitate recovery.

Beside those mentioned, another plan to which I have resorted, has generally afforded relief. It is to bathe the toe each night, pressing away the flesh from the part and inserting as much cotton as possible under the nail, and, also, pulverizing from a stick of caustic a little thereof into the corner. This, persevered in, will either relieve, or excite inflammation, in which case the old nail is destroyed, and comes off, giving you a good opportunity to correct the difficulty as the new one appears. This, though a little painful, is moderate, compared with the suffering from an operation.

If none of these plans succeed, there are two alternatives left, either to scrape the nail very thin, and apply lunar caustic upon it each day, and freely, until it contracts and separates from the flesh, or to have it pulled, cut, and twisted out by a surgeon.

The difficulty, after all, is not a growth of the nail into the flesh, but of the flesh over the nail, and, by the motion of the toe against the sole of the shoe, so as to press the flesh in position, and constant daily manipulation, not so much with the nail as the substance around it, much can be accomplished. During the period of treatment all pressure should be removed.

BOILS AND CARBUNCLES.

Job is not the only one who has repined at these unwelcome visitors, and most of us know what " as sore as a boil" means. Though usually not very serious in their results, they often last for a long time, and one crop follows another in undesirable succession. If large, the core, or hold-fast, which consists of dead flesh, is often long in separating, and if they occur close to the lower opening of the intestine, they sometimes connect with the bowel internally, and cause a fistula. At their first approach there are two ways in which you may possibly prevent their progress. The one is to take a piece of adhesive plaster and apply it very tightly around the part; the other, which is more effective, is to soak the finger in strong lye or hot turpentine. If this do not succeed, apply a leech or two, bathe the part frequently with hot water, and keep it constantly warm and moist by frequent poultices.

The ground flax-seed or slippery elm poultice are among the best that can be used. A teaspoonful of laudanum and molasses mixed, and applied to the boil before putting on the poultice, is often of service.

If the pain continues for a day or two, do not wait for the boil to head, but go to the doctor and let him cut entirely through it. Thus the pressure will be relieved, and the loss of blood is an advantage. If you have not the courage to do this, put on it a small plaster of soap and sugar mixed, as this helps to break down the skin which covers it, and over this keep a warm poultice. After it commences to discharge, you will still need poultices until pain ceases and the skin becomes whitened and wrinkled, when you may substitute a salve of equal parts of beeswax and lard, to protect it from the air. The disposition to boils is generally accompanied with the presence of too much acid in the system and derangement of the digestive organs, and they are not to be regarded as signs of full health. To prevent their recurrence, it will be well for you to use every morning as much saleratus as you can hold on a three-cent piece, or to take once each day, a half hour before eating, four or five drops of the mixture known as Lugol's solution, in a little water. It is very common to attribute boils or eruptions to bad blood, but it is not always so easy to find the cause or the remedy for this. We cannot always account for them, and rather than give a poor reason, it is better to give none at all. It is not the doctor's duty to give a reason for every thing, and it is by no means always necessary to good treatment that he should.

Carbuncles are enormous boils in which the slough, instead of being confined to the core, extends to surrounding parts. They are most common in wine-drinkers and high livers; and in old persons or broken constitutions are often very serious. The most usual situations are the back of the neck and the shoulder. They are to be known from boils by their more intense pain, by the extent of the redness and their location.

Free incisions should be made with the knife, and warm poultices applied. Where the slough is extensive, equal parts of balsam of Peru and turpentine mixed, and applied freely to it will hasten its separation. These always need the aid of a physician, for general, as well as local treatment, and though bearing the name of a beautiful gem, they are not so precious but that you may be thankful to him who enables you to dispose of them.

FELONS.

These are more appropriately named, and surely they deserve to be incarcerated. A felon is an inflammation of what is called the periosteum, or outside shell or scale of the bone. It is, therefore, deeper than a boil, or carbuncle, both of which are situated in the tissue just under the skin, and between it and the muscles. From the hard nature of the substance in which the felon begins, the abscess is very slow in reaching the surface and the surrounding flesh becomes involved, and intense pain and suffering are the result. We cannot always account for them, but sudden checking of perspiration, or rapid change of temperature, often produces them. Washer-women, or others whose hands are suddenly put into cold water, after being in warm, are especially liable to them. Boys, who go in bathing when over-heated have a difficulty of just the same character occurring on the tibia, or shinbone. When the inflammation commences, the same remedies may be used to check its progress, as were suggested for boils, but if it still continues, a free incision down to the bone is the only alternative, and this, though painful, is as nothing compared with the prolonged suffering of leaving them to make their own way to the surface. Poultices should then be used until pain and inflammation subside. If the knife is not resorted to, the best application is first leeches, then constant poulticing, and when the abscess is slow in breaking, the use of caustic, or soap and sugar over a small surface under the poultice will hasten the suppuration.

In inflammation either of the hands or feet, from any cause, or of any part of them, they should be supported by a sling, or by some other method, in a horizontal position. There are other kinds of inflammations, in which the abscess forms under thick tendon without affecting the bone, but, unable to find an exit, runs along beneath its tough covering, often to the great injury of the fingers or palm of the hand, or the bottom of the foot, or other points at which they may occur. These need precisely the same treatment as the felon. When the inflammation is located about the root of the nail, it is often so severe as to destroy the nail itself. Soft soap

5*

and poultices form the best applications, but if matter forms deeply, it should be discharged by means of the lancet. Impure washing-soda, or other irritating substances getting beneath the skin about the nail, frequently gives rise to this variety of whitlow. As to the kinds of poultices, best in all these cases, and the mode of preparing them, the chapter on that subject will inform you. The plain indication is, by renewed heat and moisture to hasten on suppuration. Narcotics, such as laudanum, hop-tea, or thorn-apple, will often alleviate pain, without interfering with other processes, and slight stimulants, such as yeast, molasses, honey, and the like, may aid in breaking down the outer covering. After discharge is established, poultices will be of service, until pain ceases, and swelling begins to subside, but should then be replaced by the application of some simple salve.

MILK ABSCESSES.

These result from a too long retention in the gland of an amount of milk greater than the capacity thereof can accommodate. This may result from cold settling in the breast so as to constrict the tubes, from neglect in removing the milk, from fever, or excitement, and from any mechanical cause which prevents the free exit of the secretion.

To prevent it, when threatened, there are two indications : first to diminish the flow to the part, and this is effected by a spare diet, especially of fluids, and by the use of a free cathartic, such as salts, thus causing free evacuation of the bowels. But a still more important point is to remove the secretion when formed. To this end the breasts should never be allowed to become over-distended. The child is the natural means of relieving it, but if this is not sufficient, artificial methods must be employed, such as the breast-pump, and the like. The external parts should, if painful, be freely, gently, and frequently bathed with an equal mixture of cold water, rectified spirits, sugar, and sweet oil, the breast supported by a bandage if disposed to drag, and the parts not overladen with covering. If the pain becomes quite severe, a half dozen leeches, followed by warm appliances, will be required. Fomentations with flannels, wrung out in hot vinegar, or spirits, or hot hop-tea, or flax-seed, or slippery-elm poultices, will often be efficacious in relieving the pain. Some one is generally to blame when a milk abscess occurs. Not unfrequently the real cause is a previous depression of the nipple by the form of dress, or a soreness thereof, so that the fluid cannot be readily withdrawn. These things should be attended to before the time for use. By slight and repeated drawing and oiling, it may be rendered prominent, and the nipple itself may be hardened by frequently bathing it with weak brandy and sugar, or if tender, by the free use of a strong solution of borax, from day to day. If it continue sore after nursing is commenced, the same applications should be used, and melted mutton tallow should be smeared over it to protect the raw points from the air. If very troublesome, the use of caustic will be necessary, and the child may have to be taught to nurse through a teat fitted over a small ring of ivory. A very strong tea of yellow dock, made thick with powdered borax and sugar, is excellent for sore nipples.

When a breast goes on to suppuration, the abscess should be opened so as to discharge freely, and not allowed to break of itself, but by relying upon the above named precautions and remedies, and not trying every thing you may hear of, the difficulty will be avoided. When, for any good cause, it is not desirable to nurse the child, the means heretofore suggested to prevent a flow of milk, and the breast-pump, should be used, and the breast penciled over with the tincture of belladonna, or bathed with a strong solution of sage leaves in gin, both of which have the tendency to dry up the secretion. In applications to the breast, care must be taken that the child, in its attempts at nursing, does not obtain from the outside of the nipple a portion of the substance applied, which may prove injurious to it.

WOUNDS.

Wounds are of varied character and extent, and the treatment depends upon their depth, or the kind of violence which the surrounding parts have suffered.

I.—A clean cut with a sharp instrument, even when deep, is not usually serious, unless a large artery or vein has been wounded, or unless it opens into some closed cavity. This class of wounds bleeds more freely

than others, and the first point of treatment is to restrain
this hemorrhage. Cold water, frequently applied, will
generally suffice to accomplish this, unless a large vessel
has been cut. If this has been an artery, the blood will be
of a florid, bright color, and flow out by jets; if a vein, it
is darker, and keeps constantly welling up. Pressure
should be made above the part, over a bone, so as to
compress the vessel. If the point of the artery whence
the flow proceeds can be seen, it should be tied with a
silken string, and no time be lost in seeking medical aid.
When the oozing from a cut part is checked, the edges
of the wound should be brought exactly together. A
graft will not grow, unless fitted accurately to the stalk,
and so the sides of a wound will unite immediately only
by bringing them in close contact. Then your business
is to keep them there. If the wound is large and gaping,
and on a part disposed to separate by the action of the
muscles, it will be necessary to make a stitch; but generally, strips of adhesive plaster will accomplish the
purpose, a small roll of which should be kept by every
family. Cut these strips broad or narrow, according to
the size of the cut, and dry the surface upon which
they are to adhere. Hold the plain side of the plaster
around a tin cup filled with boiling water, or over a hot
stove, and as soon as the plaster begins to soften, quickly
apply it crosswise over the middle portion of the cut, so
as to bring its sides in close proximity. Continue applying these strips at such distances from each other as will
bring the sides of the entire wound together, leaving
slight openings between. Then smear a little sweet oil
or lard over the part cut, to protect it from the air, and

cover it with a light linen cloth. This is the best and only local treatment required, and better than all the " terrible healing salves" of which you may hear.

II.—A bruised cut, as that made by a very dull tool or the like, needs a little different management. If it bleeds but little, use the water lukewarm, instead of cold, thus seeking to restore animation to the injured edges. Then close it as directed before, put on your adhesive plasters, and pour into it a little turpentine, or the wine and oil, as applied by the good Samaritan. Either of these exclude the air, and act as slight stimulants. If the parts become swollen and painful, it will be necessary to use the warm water dressing or poultices until the inflammation subsides, but usually these will not be required. The adhesive straps should be left on until the edges of the wound have united, when they may be soaked and carefully removed, by taking hold of each end and pulling toward the cut. If for any reason they become loosened too soon, new ones should be applied.

III.—A torn cut differs from these merely in the facts, that the edges of the wound are more mutilated, and the shreds of flesh are not so apt to unite as in a clean cut. Here again use warm water instead of cold, cut off any pieces much torn, use your turpentine freely smeared over the part, and then gently bring the sides together by adhesive plaster. Here there is not so likely to be union without suppuration, and if great pain occur, warm water dressings or poultices will be required.

IV.—A pierced wound, as from a knife, or sword or

dagger is to be treated upon the same general principles. It should not be encouraged to heal too quickly
on the surface, as if it does it will have to be reopened.
Recovery must commence from the lower part of the
wound. Where the stab has penetrated through a tendon or thick sheath, such as we find in the palm of the
hand or the sole of the foot, very severe pain sometimes
results from the inflammatory action beneath this tough
covering. In such cases poultices need to be freely
used, and it is often necessary to enlarge the opening.

V.—A gun-powder wound is a combination of the
bruised and the torn, with the addition of specks of
powder scattered about in it, acting as irritants, and
having a tendency to excite inflammation. The surface
should be cleansed, all these foreign substances picked
out as much as possible, and then the wound be treated
as those last-described. In any of these if the injury is
at all severe, surgical aid should be procured.

VI.—Sometimes a mere scratch or prick is attended
with unpleasant symptoms. The skin, as it is said, has
merely been knocked off, or a pin just entered the
flesh, and yet an inflamed and troublesome sore is the
result. In most cases all this subsequent trouble will
be avoided by covering every scratch at once with adhesive plaster, or some other oily substance. You thus
protect it from the air, and allow nature to do her work
of repair.

Soap, dirt and the like often irritate these superficial
wounds, when a little protection would avoid all trouble.
Air acts as an irritant upon all raw surfaces.

Where a wound is caused by a substance penetrating

the flesh, and remaining in it, this must, if possible, be removed, but if not, suppuration follows, and the foreign body is rejected. In rare cases it may be retained, as needles have been known to make their way along the course of muscles, and finally reappear, or sometimes, as in the case of musket balls, a sac has been formed and the ball remained without causing any further disturbance.

VI.—In case of a barbed instrument, as a spear or fish-hook, entering deeply into the flesh, there are four ways to remove it. The one most common is, to tear it out, which occasions great pain and irritation. Another is, to remove the string attached to the upper part of the spear, and pass the hook itself on through the wound ; or the barbed portion may be broken off, if it has passed through and out, and then the rest withdrawn. Still another plan, often the best, is to pass a small knife in the direction of the hook, and cut the portion of flesh and skin above it, thus permitting you to remove it easily, and converting it into a common wound.

When a wound has been made by a rusty instrument, or when, in any way, dirt of any kind has been introduced, it should at first be carefully washed, cleansed, and oiled, and a poultice applied, as soon as any decided signs of inflammation exhibit themselves.

All these varieties of wounds may, under various circumstances, need a variation in their treatment. Severe inflammation may arise from a simple wound, and if so, poultices will be required. If severe pain, laudanum may be added to the poultice, or it may be prepared with strong boiling hop-tea instead of water. When proud flesh appears, which may be known by its redness and outstanding points, it is generally a signal to stop poulticing. This, in itself, is the promise of recovery, and does but little if any hurt, so long as it does not rise above the general surface. When it does, a sprinkling of pulverized burnt alum, each day, will readily remove it.

If the sore continue discharging for a long time, and the matter from it be thin and watery, it may need an astringent, and a piece of sugar-of-lead, the size of a hickory-nut, to a pint of water, will make for it an excellent wash. Do not, however, concern yourself with every grease that is warranted to cure. Wounds often get well with them, oftener without them, and no one, so well as the discriminating physician, knows how many cures nature will make, if you will only allow her a chance.

If the methods suggested do not succeed, consult the scientific and educated doctor, who, if assistance is needed, is most likely to furnish it.

We proceed next to speak of

POISONED WOUNDS.

In these, beside the injury done to the tissue of the part, some substance is infused, which produces, or has a tendency to produce, a deleterious effect upon the system.

The bites of winged insects are among the most common inflicted, and usually are unattended with any dangerous symptoms. It occasionally occurs, however, that, either from their number, or some peculiar state of the system, serious effects follow. Persons have been known to die from the shock caused to the nervous system by the stings of bees or wasps, and even a single bite has been recorded as proving fatal. Where there is paleness and symptoms of debility, stimulants, such as brandy, wine, hot ginger tea, or the like, should be administered. The sting should, if perceptible, be removed, and turpentine, spirits of ammonia, or cologne, be applied to the wound. Indigo has appeared, in some cases, to afford prompt relief.

The bites of gnats, mosquitoes, fleas, or bugs of any kind, call for the same mode of treatment, if any is required. Strong odors are very unpleasant to all these insects, and perfumery, therefore, has some protective virtue. Hartshorn, tobacco, or any of the essential oils are efficient, both as preventives and cures.

The bites of spiders, though not so frequent, sometimes prove troublesome. and the same remedies are indicated.

It seems natural for man especially to dread the bite of the serpent, and although many of them are not poisonous, there are others whose wounds are speedily followed by direful results. The *viper* and the *rattlesnake* are the most to be dreaded, and a wound from either demands the most careful attention. A ligature should be at once applied above the wound, and the wound itself enlarged. Then ammonia or hartshorn should be applied to it, and the limb anointed with oil. Before this is done, it is well to apply suction to the wound, either by means of the mouth or a cupping-glass. It is best, first to take a little oil of some kind into the mouth, so that if the lips are sore in any place the poison may not enter. So far as the general treatment of the patient is concerned, the chief reliance must be upon stimulants. Warmth to the extremities, mustard plasters, and free administration either of brandy or ammonia, so long as any symptoms of debility are present, in frequent doses, is the great indication. Our aim must be to sustain the system, so that it may outride and overcome the depressing power of the poison. It is also important to quiet, as much as possible, the mind of the patient, for most of those thus poisoned recover, and some of those who die are seemingly frightened to death.

We next pass to notice

THE BITES OF MAD ANIMALS.

The exact nature of the poison of hydrophobia, or the reasons why it should produce such awful effects it is not within our present province to consider: but we have to do rather with the treatment of the wound. When a person suspects himself to have been bitten by a rabid animal, the very first point should be to prevent the poison from entering the general circulation. This is to be accomplished by quickness, and propriety of action. Wrap a cord or string of some kind immediately above the part wounded, and as soon as you can get a knife or a doctor, have the wound cut completely out in the shape of an inverted cone. It will then be well to apply a stick of caustic for a moment in the wound, and destroy any remaining virus. By adopting this method, you may make a troublesome sore, but its being slow to heal, will be, in this case, no disadvantage, and it will almost certainly insure immunity from the terrible consequences following the absorption of the poison. It is thought by many that the fear of evil has not unfrequently developed the disease, when otherwise the virus would have been eliminated from the system. Youatt, who has written and experimented much upon this disease, has himself been bitten several times, and has relied upon the use of the caustic alone, instead of cutting, and has passed unharmed.

It is uncertain how soon or how long after a bite the disease may be developed; but after three months the chances diminish very much. It is the opinion of most experimenters, that the poison sometimes lies for a while dormant in the wound, and therefore if not cut out or cauterized at first, it should be afterward. If at a point where cutting cannot be performed so as to remove the entire wounded part, the bite should be enlarged, and the caustic freely used. As to other methods of treatment, either for prevention or cure, physicians are not sufficiently agreed upon them for me to attempt to lay down the rules therefor. It is proper, however, in a manual like this, to name the signs by which a mad dog is to be recognized, so that care may be used to avoid any danger. I shall state the symptoms by abbreviating the particulars more fully referred to by Youatt. "The early symptoms are obscure. Sullenness, great uneasiness, dullness, shifting of posture, a strange expression of countenance, especially the eye, are among the first tokens of the malady. The dog will start with a sudden, unnatural bark, and dart at flies or small points he may see on the sides of the room. The call of his master's familiar voice will break his delirium, and for a moment he will give the familiar look of recognition. The dog, contrary to the usual belief, not only can look at, lap up and drink water and other fluids, but has an insatiable thirst." There is usually some frothing at the mouth, but not always, and the saliva is very tough and tenacious. Ellis, in his "Shepherd's Sure Guide," asserts, that however severely a mad dog is beaten, a cry is never forced from him, and Youatt confirms the remark. A mad dog usually runs awkwardly, with a staggering gait, and is disposed to snap at or bite every thing coming in his way. "In the dog I have never seen a case of madness," says Youatt, "sooner than fourteen days after the bite." Madness has sometimes resulted from the saliva upon a pimple or abraded surface. The Hon. Mrs. Duff had a French poodle, of which she was very fond, and which she was in the habit of allowing to lick her face. A small pimple on her chin was thus charged with the virus, and hydrophobia resulted, and we hope it will no longer be fashionable for ladies of distinction to patronize poodle dogs. When hydrophobia is about to show itself in a human being, the wound will usually, if still open, begin to be troublesome, and a thin discharge take place therefrom. If already scarred, it is apt to break out again. Then the person will quickly pass into a state of general nervousness. There is often restlessness, dizziness, and unusual sensitiveness. Frightful dreams, convulsive starting, suffused eyes, profuse secretion of spittle, oppressed breathing, and vomiting, are usual concomitants. Next, dryness of the throat, associated with a dislike for all fluids, betoken still more plainly the approaching trouble. Spasms, raving, a haggard countenance, a disposition to bite, and inability to drink, but add to the distress of an unquenchable thirst, and, sooner or later, death usually closes the mournful scene. Do not, I beg of you, if you have been bitten, read the description, and imagine, that because you are a little nervous, or have had a bad dream, that you have the disease; for you must remember that a learned medical professor has expressed the opinion, that out of every ten dying with it, nine are frightened into it.

As to the treatment when the disease is fully developed, we shall say but little, both because the friends ought not to trust to their own judgments, and because we have little confidence in any of the boasted methods of cure. Prussic acid, belladonna, or chloroform to control spasmodic action, and stimulants and nutritives to sustain the system, so that it may outride the terror of the storm, seem to furnish the most rational hope of relief. At the same time, the wound itself should be reopened and freely cauterized.

FAINTING.

The cause of faintness is a failure of the heart's action, and the indication is to restore this as quickly as possible. This is caused generally by sympathetic irritation, by a shock upon the nervous system transmitted to the circulation, or by a direct loss of blood. The

person fainting should forthwith be placed in the recumbent position on a bed, or, if this is not at hand, at
his length upon the floor.   Unloose the dress if tight,
and sprinkle cold water upon the face.   If it continue,
apply some sharp substance, as vinegar, hartshorn, camphor, or cologne, to the nostril, warmth to the feet, raise
a window and let fresh air into the room, and rub the
surface of the body with warm spirits.   Brandy and
water, or a little of the hartshorn dissolved in water,
may be freely administered, in case the prostration continue, either by mouth or injection.   Even if no signs
of recovery manifest themselves, the warmth should be
kept up, and medical aid procured as quickly as possible.

There is not as much danger from fainting as is often
feared, as it is very rare that the heart will not recover
its action ; but even from trivial causes it has sometimes occurred and resulted fatally, and hence none of
the means at hand should be left unapplied.   After recovery, the patient should be kept quiet for a little
time, and some nourishment provided before any exertion is made.

CONVULSIONS.

Upon their universal cause doctors still disagree, but certain facts and principles are recognized in respect to them. They consist in an involuntary spasmodic action of the muscles, and result from irritation of the nervous system. They are either general or confined to particular parts of the body. Twitching, sudden startings in sleep, grinding of the teeth against each other, and a bending of the thumb into the palm of the hand, are among the earliest and most prominent symptoms. A child falling into a convulsion should have a little cold water sprinkled or dashed forcibly into its face, should be placed in a warm bath, and a mustard plaster applied to the pit of the stomach and over the spine its whole length, to be kept on until a high degree of redness is secured. The spasmodic action should not be resisted by force, except so far as is necessary to prevent injury, inasmuch as forcible resistance but prolongs the attack. After it has subsided, if the child has eaten too hearty of raisins, smoked beef, cheese, or other indigestible substances, a mild emetic should be administered; if from teething, the gums are red and swollen, these should be relieved; if constipation, the bowels should be slightly moved, or any other functional derangement should be carefully attended to by the physician. They often occur in the rise or progress of other diseases, and are then mere symptoms of other difficulties. Spasms in a child are very often the result of the same state of system that causes delirium in the adult.

There are often cases where convulsion is not associated with any acute disease, and where, either in the child or adult, they periodically or frequently occur from slight or unknown causes. For a person to grow up subject to these, is among the greatest of earthly calamities, and no effort should be spared to overcome them. To this end, two or three suggestions may be of value.

I.—They are often perpetuated by mere habit, even after the primal cause for them has ceased to operate.

II.—Errors of diet or regimen often sustain them, when careful management might act as a preventive.

III.—They consist in a loss of the power of the will over motion, and very much can be done for them by educating the will of the child to overcome them, and by the parent assisting by an exercise of his own will over that of the child.

Cases are known, and a few have come under my own observation, where a patient could have them at will; and if so, why not prevent them in the same way? Study out for your child or friend the rules of a correct diet and management, and see them enforced. Forbid by your own stern command, their occurrence when the first symptom appears. Teach the person in the intervals, to educate his will to oppose them. When approaching, instantly dash a tumbler of cold water in the face, apply some pungent odor to the nostrils, or tickle the throat or nose with a feather.

Drinking a little salt and water sometimes seems to control or overcome them. Wearing a metallic ring, or taking an alleged specific, I have no doubt, has fre-

6

quently cured them ; not because worth itself, a straw, but because it has brought the faith and the will of the afflicted one into a state of resistance. Some experiments, which I cannot here occupy space to relate, have satisfied me how much power, mind, resolution, and sympathy, can exercise over them. There are cases, it is true, where they are dependent on some internal lesion that nothing can control ; but by following out these hints, I am sure that this lamentable disease, becoming more and more common, will often be subdued, and that, too, in a much better way than by making an apothecary's shop of the stomach. Regulation of all the functions of life, a stern will, an imagined specific, cold water, and a fond mother determined to subdue them, I have seen accomplish wonders.

HEADACHE.

Who has not had a headache, and who would not be glad to find for it a never-failing cure ? I have none to offer ; but by following a few suggestions, you may often avoid or relieve it.

Headache is usually owing to one of three causes.

1. Overtaxing the brain.
2. Disorder of the stomach.
3. Nervous affection.

For the first, the natural remedy is rest ; and, for the second, diet and regulation of the appetite.

The head should always be kept cool, and the feet

warm. Washing the head each day with cold water is
a good plan, and the feet, if inclined to coldness, may
be kept warm by a little mustard or cayenne pepper
sprinkled in the stocking.

Close fur caps, bad air, and confinement, often cause
a dull pain in the head, and the best relief is to avoid
them, and to take sufficient exercise.

In headache arising, as does the common sick headache, from derangement of the stomach, the mode of
relief is not always so immediate. A mild emetic of
ipecac, followed soon after by a stimulant, will often relieve. The best is carbonate of ammonia, or hartshorn,
about as much as can be placed on a five-cent-
piece, dissolved in half a wineglassful of water. A
seidlitz powder the next morning will complete the
work. To prevent these attacks, no food should be
eaten which is found to disagree, constipation should
be avoided, the appetite restrained, and a seidlitz
powder taken at the first premonition of an anticipated
attack, will often ward off the sickness.

There is a class of nervous headaches often arising
from causes less controllable. They are the more
chronic results of more lasting affections, or the symptoms of some derangement of the nervous system.
Bathing the head with sulphuric ether, one half diluted
with water, or the temples with equal parts of chloroform, tincture of belladonna, and of aconite, will often
soothe and allay pain, but as these taken internally are
poisonous, they must be used externally with care.
When of a plain nervous character, ammoniated tincture of valerian, compound spirits of lavender, and

spirits of camphor, equal parts, and a teaspoonful every two hours in sweetened water, will often, with due speed, restore a better state of feeling. Sometimes, when periodic, these nervous head-pains are caused by debility, and require tonics, such as your physician may indicate.

Headaches are generally *symptoms*, not diseases, and the patient who carefully studies his own case will often detect causes which might escape the less frequent glance of his medical attendant, or which, related to him, will enable him to prescribe with more efficiency.

CONSTIPATION.

Constipation, or costiveness, is so frequent a malady, and so often the cause of serious trouble in the system, that it deserves our careful consideration.

In health, the adult is entitled to a daily evacuation, and they who do not enjoy it, retain matter absolutely deleterious, a part of which, though effete, is reabsorbed into the system. It is familiar to all, that the amount voided in a case of constipation at the end of a week, though it may be larger than a single healthy evacuation, is by no means equal to what the seven would have been, and whatever it lacks of this has been returned, and had to seek its escape through the perspiration, or in some other unnatural way. Different kinds of food leave each a different proportion of residue; thus, persons on a pure meat diet, would have far less to be rejected than if on vegetable food, while a milk diet leaves very little which is not appropriated. But we are to remember that only about one-tenth of the *feces* consists of our actual food, the rest being made of the useless or spent material which has already answered the purposes of nutrition. We are, therefore, correct in saying, that it is an almost universal rule of health, that a movement should be had each day, and in small children, in whom all physical processes are more active, two or three are not unfrequent.

Let all remember, that long-continued health and habitual costiveness are sworn enemies. Like headache, it is at first, at least, a symptom rather than a disease. In one patient, it is the result of acidity of stomach, exhibiting itself by pain, acid eructations, etc.; and here a little calcined magnesia, dissolved in water, and taken from time to time, before breakfast, will usually relieve the difficulty. In many other cases, again, it is dependent upon a deficient discharge of bile into the intestines. The secretion from the liver has sometimes been termed nature's cathartic, and where this is wanting in sufficient quantity, the discharges are generally few, and scanty, and of a clayey color. A want of tone in the stomach, and a torpid state of the liver, are very apt to be associated with other symptoms of indigestion, and, in such cases, we have found equal parts of

Carbonate of Soda,
Powdered Aloes, and
Powdered Columbo root,

a very efficient relief. As much as can be holden on a

five-cent piece should be taken each morning before breakfast, until you obtain a start in the right direction, which will enable you to dispense with the use of drugs. It may be made into pills, or mixed with a flavored extract like sarsaparilla, in order to render it less unpleasant. A tea made of the common dandelion root, and drank daily, will also be found very efficacious. It may be prepared with milk and sugar, like the evening drink, and though a little bitter, will not prove unpleasant.

But the permanent cure of constipation is not to be sought in the habitual use of any medicine. This may be necessary for a time, but is only an adjuvant to more important measures. Exercise, and a proper regulation of the diet, have very much to do with overcoming the difficulty, and in vain will be all the efforts, and prescriptions of the physician, unless aided by the continuous and persistent determination of the patient. Farmers, or those accustomed to active out-door exertion are very rarely troubled with this kind of inaction.

I.—As to articles of food, there is reason to believe that some of our most usual ones are favorable to constipation. Especially is this the case with preparations of flour now in common use, and forming so large a portion of our aliment. Fine wheat flour is, to a very great extent, taken up by the system without leaving residuum, and is in its own nature binding. We believe that wheat, finely ground, without being bolted, and so, finely sifted, would be, in every respect, a better article from which to make bread, and those who accustom themselves to its use prefer it to the superfine brands. The shell or skin of the grain not only contains valuable properties not found in the kernel, but the slight roughness is favorable to digestion and proper action of the bowels, and those especially who are inclined to constipation, should give the preference to bread prepared from this.

Although in certain rare instances it may irritate, yet this can only be discovered by the actual trial.

II.—In order to avoid a constipated condition of the bowels, food must not only have bulk, but there must be a fair proportion left to be rejected; and hence a good share of such kinds of food should be used. Milk, eggs, bread, meat, and the like, as we have heretofore noticed, leave but little, while potatoes, vegetables, and fresh ripe fruit eaten before breakfast, are conducive to regularity. Certain articles of food, such as prunes, stewed rhubarb, or pie-plant, the juice of clams, etc., have a laxative effect; other foods effect special persons, so that those inclined to constipation should endeavor, as far as possible, to make their food take the place of all medicine, and, by careful management, coax the alimentary canal into working order. Where, as is often the case, there is a want of mechanical power in the muscular coat of the intestines, resort must be had to injections, and these are often of very decided service. Thus a pint of cold or luke-warm water thrown up the rectum each morning, will aid in the attempt at regularity until they can be omitted. A glass of cold water, swallowed before eating, will sometimes do for the upper bowel what the injection does for the lower. Sometimes there is a want of nervous power in the intestine, and here I have found a generous pill of hyos-cyamus, given at bedtime, an excellent laxative.

Beside all or any of these methods, the power of habit is no small auxiliary. Resolve, whether nature seems to call or not, to take time to visit the outhouse at a regular period each day, and it is astonishing how much power the habitual *will* may attain. Many a patient has been cured by the enforcement of this easy discipline. Medicines are not to be relied upon in constipation. They are only aids, themselves in a little time, by habitual use, losing their power, and our trust is in the natural tendency, aided by sensible exercise, and regulation, and such little assistance as may be required. This trust is a reliable one, and be assured, you cannot omit the due action of the main channel of the system without sooner or later incurring the penalty, and many a constitution, especially of the weaker sex, has been broken by the neglect.

Said the Scotchman to his darling son about to leave the parental roof for the duties of active life: "There are two things, my son, I wish you to remember. Keep the bo'els open ; that will do for this world. Have the fear of the Lo'ord before your eyes ; that will do for the world to come." The advice, though it may seem somewhat quaint and disconnected, was, nevertheless, very sensible.

Care as to these two points would avoid many an uncomfortable feeling in this world, and many an unhappy foreboding of the world to come. Health is no less a duty than religion, and we sin in neglecting the welfare of our bodies as well as in that of our souls.

VACCINATION.

Vaccination is the transfer to the human system of a disease known as the cow-pox, by which the small pox is either prevented or modified. In an age scarcely yet past, this latter was the scourge of individuals and nations, and thousands upon thousands suffered and died with this contagious and loathsome malady.

It having been noticed, that the disease was modified by careful previous diet and regimen, the fortunate suggestion was made, that it should be directly communicated to people purposely, after they had undergone a course of preparation, instead of leaving them to contract it in its virulent form. This gave rise to the system of inoculation in which the small pox matter was introduced directly into the flesh of those thus inoculated, and thus a modified small pox produced. Many are still living who remember the preparations made for passing through this ordeal, and the signboards warning strangers from the roads on which the doctor was distributing the virus. By this method the fatality of smallpox was very greatly diminished; but it had its objections. While some were protected, the disease, by being thus conveyed, was often contracted by the careless or unwary, and not very unfrequently death resulted to those who had submitted to the inoculation, thus leaving physician and friends to moralize upon the propriety of doing evil that good might come. It was even at this period a prevalent notion in some of the rural districts, that milkmaids who had suffered from sores on their hands, contracted from the teats of the cow, were not liable to the contagion of small-pox ; but no one before the time of the immortal Jenner had seized upon this tradition as possibly founded upon actual experience and truth. He, however, by careful inquiries and experiments satisfied himself that such protection was real, its source undoubted, and the possibility of applying it to general service worthy of a trial.

On the 17th of May, 1796, a boy eight years of age, was successfully vaccinated with the matter obtained from the cow; on the 1st of July small-pox virus was introduced into his arm, but produced no effect, thus showing, that by this simple method he was protected from this terrible contagion.   One case after another has proven the power and reality of this protection, and it has come to be regarded by the medical profession as an almost specific preventive.   It is indeed, a mild, safe and modified form of small-pox ; for, if a cow be inoculated with the matter from a smallpox pustule, a sore will form, and if from this again, matter be taken on the eighth or ninth day, and inserted in human flesh, we get a true and genuine vaccination.

The most usual point of performing vaccination upon the human subject, is upon the left arm, about one-third of the distance down from the shoulder to the elbow.

Two or three light scratches are made with a lancet or the point of a sharp pen-knife, just sufficient to start blood.   If you have a fresh pustule of another child, prick it sufficiently to allow a little of the matter to ooze out without drawing blood, and let this be introduced into the wounds thus made.   If you have the virus upon a quill, breathe upon it, and keep rubbing it immedi-

ately over the cuts made, until it shall be perfectly dis·
solved, and mingled with the blood. If you use the
scab, a piece of this about the size of a large pin-head
should be pulverized with a case-knife upon the back of
a plate or saucer, before making the incision, and when
this is made, sprinkle the powder over the scratches,
and thrust the little particles with your knife-blade into
the wound. Though the operation is a simple one, it
should, when convenient, be performed by the physician, as he can best judge whether it has taken ; but
where none can be procured, a friend may often do it
in his stead.

About the fourth day a red margin begins to surround
the point ; about the sixth day the poc is formed, consisting of a sore differing from others in its round shape,
hollowing toward the centre, and containing a fluid.
For the next three or four days the fluid becomes more
transparent, and the whole sore more distinct. The fluid
now gives place to a pus, and the scab becomes dark
and dry, and usually about the twenty-first day will
separate, leaving a depressed and indelible mark in the
skin. The perfection of the sore about the eighth or
ninth day, and not much sooner, its round shape, its fulness on the edges and depression in the centre, its pearl-
like color, and its separation after three weeks in the
form of a round scab, leaving a corresponding mark, is
the best general evidence of the genuineness of its effects.
The degree of surrounding inflammation, and the amount
of general fever, varies in different cases, and forms no
test of its success.

If it is desired to obtain vaccine virus before the scab

forms, the end of the eighth day is the proper time to take it. Prick very slightly, with a needle, the raised edge of the vesicle, and little drops of a watery-looking substance will ooze out; and if you have a few blunt ivory or goose feather tooth-picks at hand, by rubbing them over the pustule and allowing this water to dry upon them, enough will cling to each, and the effect not be interfered with.

Is vaccination really a preventive of small pox, or does it modify and lighten the disease, if contracted?

There can be no doubt of it. A great majority of those once properly vaccinated, will never contract it at all, and those that do, will generally have it less severely. A careful investigation was made of this matter in our own country, some fifteen years since, and a series of carefully collected facts proved that, in the unvaccinated, small pox is five times more fatal than in those having varioloid, and that perfect vaccination protects as completely from small pox at all, as having had it once does from having it a second time. Statistics in other countries have proven the same thing. In Boston, where vaccination is carefully and generally performed, there was but one death from small pox or varioloid, during the whole of last year.

Should vaccination be repeated? A majority of those once protected, ever remain so. That it does run out in some, and that it does not run out in others, has been repeatedly shown; and there is no other way of determining whether or not it has, in any particular case, except by re-vaccination. An important change of life, as at the period of puberty, marked changes of climate, severe attacks of sickness, or other eruptive disorders at the time of vaccination, sometimes seem to modify its effects.

After a careful comparison of facts, we would suggest the following rules:

I.—Let every one be once vaccinated. The parent who neglects it is to blame. He knows not when his child may be exposed to a horrible disease, from which he has the power to protect him. If there is any doubt as to its having taken, or as to the mark being genuine, let it be immediately repeated with some different matter, and thus you may, at the time, test it. The best time for vaccinating is, after the third month and before the seventh; that is, before the time of teething. If neglected at this time, any other period when the child is in good health, may be chosen. The means of vaccination are cheaply within the reach of all, and as it is far more neglected in this country than in most others, there is need of reform. The public good requires its more general application.

II.—Re-vaccination is proper after the child has passed the period of boy or girlhood, and at least in all cases where the person is directly exposed, ought to be repeated by fifteen. In many of the standing armies of Europe, both the value and need of re-vaccination have been tested. In these, small pox was formerly most fatal, and its ravages most extensive; but by a rigid legal system of re-vaccination, it has become almost annihilated. With a careful regard to their personal and the public good, on the part of parents, physicians, and the people at large, we see no reason why small pox should not become well nigh eradicated from our land.

When parents do not know what aileth their children, they have a strange fancy for guessing that they must be troubled with worms. That such is the case is sometimes undoubtedly true, but nothing is more unsafe than upon this mere supposition, to commence giving them some of the severe patent vermifuges. From their use I have known more than one child to be thrown into violent convulsions; and I remember well a bright little boy who was for a long time paralyzed in the lower limbs by taking, according to *directions*, one of the popular worm-killers of the day.

There are various kinds of worms found in the human intestines. People who have nothing else to think of, sometimes imagine they have a tape-worm which is about twenty-five feet long, and about one in a thousand of those thus suspecting it, really have one of these lengthy animals; but they need so much the charge of an attending physician, and the symptoms betokening them are so variable, that I shall not describe them here.

The two more usual varieties are the long,·round worm, and the thread, or pin-worm. The first usually occupy the upper part of the bowels, and the latter the lower.

The following are the chief signs of the presence of worms: Wandering pricking pains in the abdomen, generally relieved by eating or a draught of cold water; variable appetite, and alternate diarrhea and costiveness; pallid countenance; itching nose; offensive breath; and attacks, so to speak, of general uneasiness.

With the thread-worm, these symptoms are not so distinct, but as they usually occupy the rectum, there is, beside, excessive itching of the anus or lower part of the bowels.

We cannot always assign a cause for the presence of worms, but we know that children reared on raw or coarse vegetables, without much meat, or those freely using sugar and cheese, or abstaining from salt, are most liable to them.

For removing them there are many much praised remedies, but I shall refer only to a few.

Against the round-worm, spirits of turpentine is one of the best and most effectual. It may be given in the morning, *before breakfast*, with a little peppermint, brandy, honey or milk. A child of twelve years may take a small teaspoonful, one of four, a halfteaspoonful, and in proportion for different ages, to be followed up by a dose of castor oil in four hours, if it does not ope rate. The seeds of the Jerusalem oak are also quite efficacious, and are frequently used. These may be given after being slightly bruised, to a child four years old, in halfteaspoonful doses, twice a day, and after three or four days, followed up by a mild cathartic. They are best administered in a little molasses.

For the small thread-worm, the best remedies are injections. A tablespoonful of common salt, in a teacup of warm water, thus given to a child, each morning or night, for two or three days, will often have the desired effect. If the itching still continue, an injection of fifteen drops of camphor, and half a teaspoonful of turpentine, in double the quantity of sweet oil, for a child of four years, will be still more efficacious.

How shall the recurrence of worms be prevented? Good food, exercise, and fresh air, are all important; Vegetable, such as turnips and cabbage, beets, and all unripe fruits, are objectionable. Common salt, in half teaspoonful doses every morning, for a child under six, and in teaspoonful doses, if older, is one of the cheapest, best, and readiest of anthelmintics. Where there is much paleness and feebleness, an iron tonic will be indicated.

Food should be carefully masticated and eaten regularly, and acidity of stomach should be corrected by a little lime-water or pearlash; a little salt given each day, and permanent relief and exemption will usually be secured.

SPRUE.

The Thrush, or Sprue, as it is more commonly called, is a disease of infancy, and generally, I believe, caused by the carelessness, of the doctor, mother, or nurse.

It arises from irritation of the mucous membrane of the stomach or mouth, caused by sour food, and by particles of milk remaining in the mouth. To prevent it, the child should be nursed or fed so as not to overload its stomach, and if the milk is rejected in a curdled state, one quarter of a teaspoonful of lime-water may be given the child each day. This is prepared by throwing

a piece of lime, or a little common farm-lime, unslacked,
into a small quantity of water, and, when the lime
settles, you have lime-water, which, if you put enough
lime in, is always the same strength; for the water will
take up or dissolve only just so much. After nursing,
or, at least, each morning at its washing, the child
should have the mouth cleansed with a rag wet with
cold water, and passed around the mouth over the
finger of the nurse, and by thus removing all particles
of food, and avoiding acid stomach, you will avoid
sprue. When it has already made its appearance, the
best remedies are, a little magnesia or prepared chalk,
dusted freely in the mouth after cleansing it, and if not
soon controlled, use each day a solution of borax, strong
enough for you to perceive it by the taste. These suggestions followed out will prevent most cases and cure
the remainder.

MUMPS.

This is one of the contagious diseases which is seldom
serious in its character, but in its approach often excites
much anxiety in those fond of eating and talking. It
consists in a swelling of the gland known as the parotid,
just below the ear, and those adjacent to it. This, as it
enlarges, stiffens the jaw, so that the mouth can scarcely
be opened, raises the point of the ear, and, by the ex-
tension of the face in breadth, imparts to the countenance a somewhat ludicrous expression. Sometimes
only one side is affected, and at others both at once, or

in succession.   There is generally some fever, and
locally the parts are painful and tender.   In favorable
cases it reaches its height by the fourth day, and then
gradually declines.   If communicated to others, they
will perceive it about ten days from exposure.   Sometimes the disease will suddenly disappear or abate, and
show itself in some other gland, as the breast in the
female, or the testicle in the male, and now and then we
have inflammation of the brain following.   This, however, is very rare, and it is usually a mild disease.

The treatment consists in evacuating the bowels by a
small dose of salts if they are constipated, and in employing warm applications locally.   Warm sweet oil
and laudanum rubbed upon the swelling, and dry, hot
flannels applied over it, generally suffice; but, if there
is much pain, bags of hops steeped in vinegar may be
frequently applied.

When the disease changes locality, or is accompanied
with swelling in other parts, these same appliances
should be continued about the neck, quietude observed,
and no interference will be needed unless the symptoms
are severe.   If so, or if there is decided tendency to
the brain, indicated by continuous delirium and high
fever, no time should be lost in procuring medical aid.
We have sometimes known the patient, especially if an
adult, to be left in a nervous excitable state, but this
will be overcome by time or the use of mild tonics.

CHILLS AND FEVER.

In many sections of our country this is one of the most common forms of disease. In popular language, chill is the term used when there is merely a sense of coldness; ague, when this amounts to a continuous shivering, and intermittent fever, when the attacks have only short remissions; but, in fact, these all are but slightly modified forms of the same affection.

Not a few, from their own experience, know something of these

> " Chills that set the bones to aching,
> Giving them an earthquake shaking,
> Causing every tooth to chatter
> Like bones shaken on a platter,
> Twisting all the joints about
> With a wrench that makes you shout."

And many a one is ready to add—

> " Climax of all earthly ills
> Are these racking fever-chills."

A single chill may arise from various causes, such as exposure to sudden changes, disorder of the stomach, chronic disease, the formation of abscesses, the advent of inflammatory action, and the like; but in these cases the cause is different, and there is generally but the one attack, followed by the symptoms of some special affection, thus heralding its approach. Chills and fever, on the other hand, is a periodic disease, having a tendency to return usually the next day, or the day after that, at nearly the same time—sometimes the return is every fourth day, and cases are on record where they have occurred at the rate of two daily.

The attack divides itself into three stages; first, the chill, the character of which is portrayed by the word, and the experience, or the sight of one, better than by pages of description; this followed, in an hour or two, by flashes of heat, intense thirst, and fever, and this, in turn, passing into a copious perspiration, leaving the sufferer weary and debilitated. In some cases, between the attacks, the appetite is good, and the feelings of health seem struggling to return, but another chill soon sets hope to quivering, and thus alternate depression and relief try the constitution, the temper, the spirits, of the patient and are even worse than more alarming ailments. While the "ague is upon him," hot coffee, or other warm drinks, may be given, and external warmth applied in every way; but there is not much else to be done, unless symptoms of congestion or prostration occur, which demand medical aid. Where there is great sinking, active stimulants, external and internal, are required, and if there is stupor, mustard, and other active counter-irritants, must be used; but it is rare that there is any immediate danger. During the succeeding fever, cold water is both grateful and useful, in promoting perspiration, and cold to the head, a teaspoonful of valerian, and warmth to the feet will aid in relieving the intense headache. When sweating supervenes, draughts of air should be avoided, the surface frequently dried with flannel, and the patient be kept quiet.

Among the multitude of remedies for chills and fever, Peruvian bark, in some of its forms, still holds the preeminence. This is obtained from a tree growing in South America, and was first introduced into Europe as a medicine by the Count of Cinchon, whose wife, during a residence in Peru, was cured of an ague by its use. It once sold for its weight in gold, and although its employment was alternately approved and condemned, popes recommending and kings opposing it, yet its value has been proven by the still higher test of time, and in the varied forms of quinine, chinoidine, China, prepared bark, and the like, it is still the main reliance even of those who profess not to use it. Besides this, charcoal, piperine, arsenic, willow and dogwood bark, gunpowder, and various other remedies, are employed sometimes with manifest advantage, and after-effects often attributed to some of the articles used, often result directly from the disease itself.

Our chief design in speaking of this affection, is not so much to designate the precise method of treatment, which must vary with particular symptoms, as to give such general directions as are equally important in preventing attack, avoiding improper treatment, and expediting and securing permanent relief.

I.—It is not generally safe to take any tonic or anti-periodic medicine, immediately after the first chill, in a first attack. It is not easy for an unprofessional person to know that the disease is of a periodic character, and if it is the signal of remittent fever, erysipelas, or inflammation of the lungs, the tonic would be injurious. It is better to await a single recurrence, rather than run the risk.

II.—A free cathartic is usually requisite in a first

attack, before other remedies are employed. A good dose of antibilious pills, not patented, or of other search-ing medicine, will often do much toward expediting cure.

III.—A patient who has been relieved, should always remember that the disease is decidedly liable to recur, especially so after one, two or three weeks, and hence, from the fourth up to the twenty-second day after the chill is broken, some mild tonic should be taken daily.

IV.—The disease is miasmatic, i. e., is always first caused by a specific poison existing in the air. This is confined to special localities, and is especially prevalent in the vicinity of marshes, or where, by the woodman's axe or the upturning plow, new, rich lands are exposed to the sun. In its effects it is especially active at morning and night, loves the ground, and the foliage of trees, and is most prevalent when heavy rains are succeeded by intense, continuous heat.

Hence it is readily inferred that avoiding exposure to morning and evening dews, sleeping in upper apart-ments, and sometimes even change of locality, are im-portant methods of relief. Bishop Heber, in his narra-tive of Northern India, states that during the summer months, not only men, but all animals and birds, flee from the malarious woody districts to the open plain.

V.—It is worthy of note that travelers through a miasmatic district, or late settlers, will be affected, when others are not. So much is this the case in our own country that severe forms of the disease are often known as the "stranger's fever," and hence, in inquir-

ing into the health of such locality, that of old resident-
ers is not to be taken as a guide.

VI.—Those who have once suffered an attack are, for
a long time afterward, more liable to a recurrence. In
this particular it is the opposite of most affections.
When, however, the original effect is overcome, a
degree of protection is acquired. A friend who contracted an attack in passing the Pontine marshes in
Italy for seven successive years, at nearly the same
season, suffered relapse ; and abundant cases are recorded exhibiting this marked periodic tendency.

VII.—Over-fatigue, derangement of the system, imperfect clothing, and imprudent exposure, predispose to
the disease. Where persons are compelled to expose
themselves to miasmatic influences, care should always
be exercised to have the system in a healthy state, and
it is often well to use some mild tonic, such as the
citrate of iron and quinine, as a preventive. In conclusion, we may add that nothing so certainly undermines
the constitution as continuous chills, and if medicine
and due care fail to prevent or remove them, another
locality must be sought, for all other temporal induce-
ments are not to be compared with the blessing of
health.

This is a spasmodic affection, commencing like a common cold, but soon the cough is accompanied with the characteristic hoop. The duration of the disease varies very much in different epidemics, and is affected by slight circumstances. In mild cases it usually attains to its height by the end of the fifth week, but, even after it begins to abate, slight cold or general derangement of the system will renew it in all its former severity. A loud distinct hoop is always a good sign, for in the worst cases the suffocative character is so marked that air does not reach far enough down into the lungs during the spasmodic action to produce a full " whoop." Hooping cough is chiefly dangerous from its complications, such as bronchitis, inflammation of the lungs, diarrhea, convulsions, or nervous difficulties, which are apt to be developed with it. The average duration of the disease is about ten weeks in summer, and twelve in winter; but sometimes, by aggravating circumstances, it will continue much longer. It is a contagious disease, but how long it retains this character, it is difficult to determine. It is believed, at least, to do so throughout its usual three months duration. Very small infants, with a little care, will rarely contract it.

Hooping cough we regard as an ailment over which medicine has considerable control, and when met early in its progress its symptoms may be checked and modified, and sometimes its duration shortened. The following preparation is the best general remedy, but if serious complications occur, the direction of the physician will be needed:

> Syrup of Squills,
> Syrup of Ipecac,
> Syrup of Tolu,
> Purified Linseed Oil,

of each one ounce, mixed. Dose for a child of one year, one-half a teaspoonful three times a day, and of two years and over, a teaspoonful. Where there is great oppression, it may be given every half hour until vomiting results. A little flax-seed tea, with a small teaspoonful of saleratus to the pint, is very well as a frequent drink for a child. Good linseed-oil alone is a valuable remedy in hooping cough, and its taste may be improved by a little winter-green or vanilla flavoring. Hydrocyanic acid is often of great service in controlling the spasmodic cough, especially if there is a tendency to convulsions. This may be added to the mixture above named, but should only be used under the direction of the doctor. It is well, also, to make slight counter-irritation over the chest. An occasional mustard-plaster, bathing the chest with opodeldoc or turpentine, or some other slightly irritating liniment, or a brown paper greased and sprinkled over with ginger, will often be of some service. Exposure to a damp atmosphere should be avoided, the skin be kept in a healthy state by an occasional warm bathing, and its surface well protected by flannel from sudden changes of temperature. There are a host of other popular remedies, but they who try the whole range, from cochineal to gunpowder,

7

will do no better than to trust to those already mentioned, and to seek medical aid if complications occur.

BLISTERING AND BLISTERS.

Blistering is often an effectual means of subduing internal inflammation, or of relieving internal pain and congestion, and hence is often resorted to in the treatment of disease. It is especially applicable in chronic disease, or in acute, after the active inflammatory tendency has been subdued, and is frequently ordered by the physician, when he cannot remain to attend to its application or future management.

A blister on the skin, which consists in the raising of the cuticle or scarf skin, may be made of any thing which severely irritates the surface, as pepper, ginger, croton oil, mustard, spirits of hartshorn, and the like; but usually, when the full action of a blister is desired, the Spanish fly ointment is the best article to be employed. This may be spread on a piece of chamois skin, or kid, or leather, or factory muslin, or brown paper, of the size desired to be applied.

It may be spread evenly with the blade of a case-knife, or with the ball of the thumb, moistened to prevent its sticking thereto. Let it be about one-eighth of an inch in thickness, and not much warmed, as heat injures its power. It should be applied to the part of sound flesh nearest over the seat of pain, or of the disease. The surface should be bathed and washed with a little warm water. If there are follicles of hair, as on the calf of the leg, these should be shaven off, and then the blister smoothly applied, and made to press slightly against the skin, by a bandage over it. It will usually need to remain upon a grown person from ten to twelve hours, but no precise rule as to time can be given. After a few hours, especially if the patient complain of it, you may lift up one edge, to see if the cuticle or skin is raised, and if already loosened and beginning to fill up, you may at once remove the blister. From children and very delicate persons, blisters should usually be removed as soon as the very first signs of raising occur. In taking them off, great care should always be exercised not to tear the raised skin, and not to leave any of the blistering salve adherent to it, for if you do, this is very likely to affect the bladder, and cause great pain with retention of urine. To be certain of avoiding this, it is often well to place over the blister, before putting it on, some very thin paper, or book muslin, which, by being tightly pressed, will not prevent the action of the plaster through it. Sprinkling a little powdered camphor over the blister will also prevent its affecting the urine, and not interfere with its local action. As soon as the raised cuticle fills with water at the lowermost points, it should be cut or snipped just enough to allow the fluid to ooze out, and this is best of all dressings, for it will settle down upon the raw surface, and you may then cover it with a little sweet oil, or wilted cabbage leaf, or a salve made of equal parts of beeswax and lard melted, and smeared on a piece of linen, with powdered chalk, a slippery elm poultice, or any thing else which will not do much but suffice to keep it from the air. Poultices or stimulating salves are best, when you desire that it shall not heal very speedily. They generally heal without much trouble, but if very sore, after a warm poultice to subdue any inflammatory action, a salve of fresh butter or cream and prepared chalk, equal parts, will promote recovery. If they come to be troublesome ulcers, they ought to be treated according to the directions you will find under that head.

SORES OR ULCERS.

What is good for a sore? is a very common question, and one for which many unprofessional persons have a very ready answer. You might just as well inquire what is good for sickness, and pretend to prescribe for every ailment alike, as to seek to treat all sores precisely the same. They differ very materially from each other, and must be met by corresponding varieties of treatment.

I.—We notice first, what is known as the *common healthy* ulcer. These are distinguished by the bright arterial color of the surface, by an even whitish edge instead of a thick, dusky, raised one, and a light cream-colored discharge of pus. The treatment of such an ulcer is very plain and simple. Locally, the sore should be cleansed each day with castile soap and warm water, and if without pain, no application is needed except a little oil or soft plaster to protect it from the air. If situated on one of the limbs it is well to bandage the part from below, and to keep the limb in a horizontal position. The state of the general system often needs some attention. If the tongue is coated or the bowels constipated, a teaspoonful of sulphur and cream of tartar equally mixed and given each morning, will be of service. With attention to both the constitution and the sore, recovery is usually speedy. When an ulcer gets proud flesh in it, the small red points elevate themselves above the surface, or bleed easily, and all that you need do is, to sprinkle each day a little powdered burnt alum upon them until they are removed.

II.—Next, we refer to the *Indolent ulcer*, or that which is oftener known by the name of an old sore. In these there is a deficiency of action. The edges are thickened, uneven, and of an unnatural color, the discharge thin and a little watery, and no appearance of proud flesh or healing. Such a sore as this stands in need of stimulation. Wash it with spirits and water, and if this does not produce any smarting, add a little saleratus thereto. The part affected should be kept horizontally, and carefully and evenly bandaged. Red precipitate ointment as a dressing, for a few days, is often of great service. If the ulcer have a dark, rough, dead appearance, a yeast poultice, or one made by pouring strong, hot sassafras tea on flax-seed, instead of hot water, will prove of advantage.

Great attention must be paid to the condition of the system. The digestive organs must be put in good working condition, the bowels regulated, and then food such as meat, milk, soft-boiled eggs, and good bread and butter will be of benefit. To improve the condition of the stomach, you may choose any one of the following medicines, to be followed up until the tongue assumes its natural healthy appearance: Pulverized charcoal and magnesia, equally mixed, and a teaspoonful in milk, before breakfast, a blue pill every other night, followed by a half-teaspoonful of powdered rhubarb in the morning, or a teaspoonful of salts and cream of tartar, or sulphur and cream of tartar every other morning. Any of these taken moderately will, in general, after a little, improve the tone of the stomach, and prepare it for the stimulus of food—the best purifier of the blood.

III.—Ulcers that are inflamed, irritable, and painful, often stand in need of poultices and anodynes. By dressing them with a poppy-leaf or hop poultice at night, and applying prepared chalk during the day, they will generally change to the common ulcer. A Dover's powder, or a few drops of laudanum, internally, often aids in these cases; for the whole system not unfrequently sympathizes in the irritation.

IV.—There is a variety of ulcer known as the *varicose*, which often proves very troublesome. This is dependent upon an enlargement of the superficial veins, and usually occurs on the leg. These become so distended as often to burst, and thus form a sore very difficult to heal. As soon as any enlargement of the veins is noticed, especially if there be a dusky appearance of the skin, the part affected should be washed each day with salt and water, and a bandage applied, or a lace stocking worn, so as to prevent the blood from settling in the part. If the ulcer forms, the same mode of treatment is to be adopted, the sore to be kept cleansed, and slight pressure to be made upon it, by means of lint, with strips of adhesive plaster over it. Such are the general points in respect to the treatment of the various varieties of ulcers, and without you are able definitely to determine their character, it is best to apply to some one who, from knowledge and experience, ought to understand their management better than yourself. What we have thus briefly said of them is sufficient to exhibit their variations; to show that they are often not only local but constitutional evils, requiring careful general treatment, and when such is the case, we should not trust to our own judgment or medication.

DEFORMITIES.

It is not our intention or design to mark out the modes of cure in each of these, but rather to draw the attention of parents to the importance of promptly seeking the means of relief.

Club-foot, hare-lip, nævi, or mother marks, as they are generally called, and other malformations, as well as results of injuries and accidents, can often be effectually cured or relieved, if attended to at the right time.

It is sorrowful and sinful to have a child grow up with any disfigurement which might have been easily relieved in infancy, and whenever we meet such a person, we feel a pity that kind friends should not, before growth was complete, have sought the desirable relief. The best time usually for operations of this kind is from the third to the sixth month, and if not attended to so early, the child should not be allowed to grow very old before the operation is performed. It sometimes seems hard to wound an infant, but it occasions them less pain, and, from the yielding and growing nature of the parts, the operation is more completely successful than if postponed to later years. Let the case be placed, not in the charge of any pretender, but of one who is recognized by those of the same profession as an adept in his art. In nothing is the success and triumph of surgery so fully displayed, as in some of these plastic operations which supply defects, and restore beauty in the place of deformity.

FOREIGN BODIES IN THE EYE, EAR, OR NOSE.

The eye is of necessity exceedingly sensitive to the presence of any foreign body, and unless it is speedily removed, inflammation must ensue.

Where any substance is thrust into the eye, and sticks by its sharp point into the outer coat, it is very important that the eye itself be kept perfectly at rest until this can be removed. If the substance can be seen, it may be taken hold of and quickly removed by a narrow pointed forceps; but if not, the eye should be kept closed until a physician can be procured. Unskillful attempts at extraction only aggravate the difficulty.

Where sand, gravel, or any thing else has got in the eye, between the lid and the ball, the first thing done should be to close the eyelids and keep them so for a little time. The advantage of this is obvious, on a moment's reflection. The foreign body causes a copious effusion of tears, and then when you open the eye there is present a little stream of water to wash it out, and wash away the offending substance. If, on the other hand, you keep winking and blinking, there is no chance for the tears to accumulate, and the irritation is increased. If persons would have their thoughts about them when any thing is got into the eye, and pursue a judicious course, the foreign body would be much more likely to come out at once, than it is by rubbing and working at it. But if you cannot thus succeed in ridding yourself of your annoyance, let a friend press down the lower lid, so as to see its inside surface, and the upper one may be easily everted by taking hold of the eyelashes; then at the same time pressing the upper part of the lid down by means of a little stick like a pen handle, and raising the lower part up, thus turning the lid quite completely, and enabling you to view its whole surface, and if any speck is seen, it is easily removed by the twisted point of a silken handkerchief, or any other soft material. Dropping into the eye a drop of weak sugar of lead water, or of cold green tea, two or three times a day, will be of service, if there is much inflammation. If there is considerable pain, soak the eye with warm water and laudanum, before retiring. If it is not soon relieved, seek for other remedies, for, the light of the body is the eye. Diseases of the eye are sometimes very in-

7*

sidious in their approach, and speedy in the destruction of the powers of vision. If a young infant has sore eyes, they should be carefully watched, and protected from the light, as there is a disease of them common to small children, very serious and rapid in its character. Any inflammation of the eye which causes the pupil to look dusky, so that a person standing before it cannot plainly see his image, should command early and watchful attention.

Foreign substances in the ear are not easily reached, and should be worked at with the greatest caution. If a bug or insect enters the ear, a little sweet oil, or two or three drops of turpentine mixed with an equal quantity of castor oil, will be very likely to induce him to return. The ear should be syringed out afterward, with a weak solution of saleratus water, until the oil is removed. Where solid substance has entered the external ear, effort should be made to get behind it with a slender bent wire, or slightly curved ear instrument; and if this do not succeed the ear may be expanded by an instrument prepared for that purpose, and a small forceps will enable the operator to remove the substance. No severe efforts should, however, be made, for often nature will in some way get rid of it, and inflammatory action is not always excited even when the foreign body remains some days.

When a foreign body lodges in the nose, or gets there by the somewhat favorite amusement of children, a full inspiration of air should be taken through the mouth. Let one nostril be closed, and the air blown quickly

·through the one in which the article may be lodged. If this plan do not succeed, a little snuff will often promote sneezing sufficient to remove the offending article. If, after repeated trials, these methods do not avail, the same must be resorted to as in the case of a solid substance in the ear. When it cannot be extracted, it is best to let it alone, if it occasions no great uneasiness, and often some accidental effort will dislodge it. If it prove troublesome, it may be pushed still further up, until it falls into the throat, and thus is removed.

CHOKING AND SUFFOCATION.

These are so intimately connected with each other that we might, with propriety, consider them in connection. As they have to do with the throat and lungs it will be necessary to glance, for a moment, at the relation these bear to each other. The gullet, by means of which food passes into the stomach, is situated behind the windpipe, and, unlike it, in a state of rest, is partially closed; for, being a loose yielding bag or tube, it is only at full size when distended. Not so, however, with the windpipe. This is needed for constant service, and air is so necessary every moment, that it is very important that it should have a firm, open passage to the lungs, and so we have the windpipe a large tube kept open and firm by the rings of cartilage you feel when you press your hand against the throat. These do not extend around so firmly to the back part of the throat, and thus, in the swallowing of food, this gives a little to make room therefor. In order that food and air may go each to its appropriate place of destination, nature has provided us, in the upper part of the throat, with a beautiful and effectual arrangement. By standing before a mirror, and pressing down the tongue, you may easily see a valve-like cartilage, with a yielding edge, projecting out from the region of the windpipe. A plain man once asked a pretended doctor what that little clapper was for. He told him that it was to separate the fluids from the solids, that food went down the one passage and drinks the other, and that this little valve had the power of determining which course each should take. "Oh!" said he, "how that must jump when I eat mush and milk." My reader, however, will not need to be informed that such is not its use, but when food or drink is swallowed, the part directly beneath this valve rises up against and adjusts itself to the precise shape of the organ, and thus a perfect closure of the upper orifice of the windpipe is produced, over which the substances swallowed slip, on their way into the stomach. When no food is swallowed, the passage is constantly open, and the air easily and naturally flows in peaceful current to and fro, down to the minute air-cells which make up the lungs.

Sometimes, by too rapid eating, or by some purely accidental position or circumstance, a little water, or a crumb of bread, or some larger and less yielding substance, will find its way between this valve and the windpipe, and the person is said to be strangled. Or a piece of food may be swallowed, and, on account of size, shape, or some other circumstance, may become fast in the gullet, thus pressing against the windpipe, on the side adjacent to it, and causing a feeling of stricture, more or less severe; or a rope may be drawn about the neck so tightly as to press upon the windpipe; or, if foul or very impure air enters the mouth, it causes a sudden spasmodic action of the throat, so that the upper part of the windpipe closes itself, to prevent the entrance of the noxious air, or, if any of it does find its way to the lungs, it but increases the difficulty.

Again, in drowning, by direct mechanical means air is prevented getting into the mouth. We give different names to these modes of interference with the natural process of respiration, but the result in all of them is the same, namely, stifling or suffocation. The object of each breath is to purify the blood as it comes to the lungs loaded with the carbonic acid and other noxious materials, which it has been its business to collect as refuse matter, in its course through the system, and to furnish in its place the pure atmosphere by which chemical changes are wrought. Interfere with this indispensable process by any method, and you have a suspension of life-power either partial or complete.

With this understanding of what suffocation really is, we are prepared to notice the various impediments to natural breathing which may occur.

I.—*Things in the Windpipe.*

When a foreign substance, from any cause, gets into the windpipe, there is at once a violent spasmodic action of the throat, by which the valve-opening is closed, and as long as the foreign body remains, it tends, both by its actual presence and by the irritation resulting, to interfere with the breathing. When, as the expression is, any thing has thus "gone down the wrong way," the first movement is to press upon the chest, and make an attempt at a sharp cough. If the substance is not fast or heavy, this often will dislodge it and cast it back into the throat. Thrusting the finger down to, and beneath the edge of the epiglottis is often efficacious, for sometimes the article is just over or near it, and can be got at with the finger, and if not, the reversed spasmodic action caused in withdrawing the finger is favorable to expulsion. Accidents of this kind are attended with very variable results. If the article thus passing into the windpipe makes its way far down, the irritation is usually less than where it remains near the epiglottis, and in some cases substances have remained for life' without causing trouble; in others they have been rejected by coughing, after months or years. In some instances, again, they have given rise to actual inflammation or chronic disease of the voice-organs and lungs,. while often they have been the cause of speedy death.

A case is on record, where a man who while holding a bullet between his teeth, let it slip so that it rolled into the windpipe, but had it dislodged while suspended by his feet, the weight of it starting it from its unwelcome position, and causing it to roll out again ; but so fortunate a result could not usually be predicated. Where strangling continues after the first effort to overcome it, a surgeon must be summoned in haste, that, if other means fail he may afford relief by opening the windpipe. Where death is already threatening, and no professional aid can be obtained, we think that even a friend would be justified in making, with the point of a knife, a small incision into the windpipe, lengthwise, just below the Adam's apple, as it is called, thus giving the patient a better chance of life.

As this is a kind of accident very serious in its nature, the notice of it should be a warning to parents, children and all, not to talk rapidly or impassionately, or laugh immoderately with the mouth filled with food, and never patronize the mouth as a repository for pits, pebble-stones, buttons, pins, or any other articles upon which neither lungs or stomach have any demand.

II.—*Articles fast in the œsophagus or gullet.*

When any substance has fastened itself in the gullet, the symptoms are not usually so rapidly severe, but the pressure upon the windpipe causes great discomfort, and may, even, result in suffocation. Where it is a sharp substance, as a splinter of bone, it may not only affect the breathing, but may cause such irritation as to result in inflammation and death. If the thing swallowed is within the reach of the finger, it should, of course, be extracted, and nature so quickly prompts us to make the attempt, that much instruction is not needed on this point. If this do not succeed, swallowing some stiff food, as bread or mashed potatoes, and the drinking freely of water, may be tried, in order to force down the article, but this should be done somewhat cautiously, for if it do not succeed in removing it, it but increases the bulk and pressure. If still the trouble remains, an emetic of ipecac, or sulphate of copper, may be given, or of ground mustard, a teaspoonful in warm water,

every ten minutes, until vomiting is excited, in the hope
of thus dislodging the evil. If none of these methods
succeed, and the symptoms are still urgent, the substance must be pushed downward, by means of a pro-
bang, or a hickory stick well wrapped about with soft
material, as India rubber or linen, at its point, and
pressed with some force directly down the gullet. A
happier old lady I have never seen, than one who, having hastily swallowed a piece of meat, found herself on
the verge of suffocation and death, but was finally relieved by this process, and she is very confident, tooth-
less though she be, that she will never repeat the experiment. Where such articles as pins or pennies have
been swallowed, unless there are serious symptoms, too
active measures should not be at once resorted to.
Even after the article has passed into the stomach, there
is often a pricking sensation, from the irritation in the
gullet, as if it was still there, and, when in the stomach,
it is best to give a little sweet oil, and solid food only,
and not resort to cathartics too soon, as these but make
more fluid the contents of the bowels, and thus the real
trouble is often left behind.

III.—*Suffocation by hanging.*

There are but two classes of persons who generally
die from this cause. Those who deserve it, and those
who desire it. For the first, we seldom have the opportunity or inclination to use any curative measure ; and
the second class are becoming much less numerous than
formerly, more scientific methods of suicide being now
preferred. In hanging, death usually ensues from suffocation, caused by the direct mechanical pressure of the

cord upon the windpipe, preventing the circulation of the air through it, but sometimes it is hastened by the pressure upon the large veins of the neck, the circulation of the blood thus, also, being impeded.

The first indication is, at once to cut the rope, and thus remove the primary cause. Not a moment should be lost to untie a knot or give alarm, for in such a case, seconds count. Sprinkle cold water suddenly in the face and over the chest, let in plenty of fresh, pure air, apply warmth to the extremities by means of hot bricks, bottles of warm water, rub the surface with warm spirits or whatever stimulant is at hand, and in other respects follow the general treatment you will find herein recommended in the management of the drowned. Although comparatively few of the hanged are recovered, yet no pains should be spared, for success is sometimes obtained, a human being relieved from the most dreadful of deaths, and not unfrequently the attempted suicide is abundantly thankful for his deliverance from his own cruelty. While these efforts are being used, the surgeon should be on his way, that such other means as his judgment and knowledge may suggest shall be rendered available.

IV.—*Suffocation by foul air.*

This is among the most common means of suffocation and its causes, and the remedies should be well understood. We have already remarked, that necessary changes in the blood, and consequently life depends upon pure air being furnished to the lungs. Now, foul air produces two difficulties: by being breathed into the lungs it acts as a direct actual poison, and by its irritation when in any large quantity, it causes a *closure* of the opening at the epiglottis, thus preventing the entrance of any air whatever. In either case, if not relieved, suffocation and death is inevitable.

Carbonic acid is the most common of the gases deleterious to life, and hence requires the most special notice. A trace of it is constantly present in the atmosphere, it is given off by every breathing animal, absorbed by every leaf, is one of the constant results of all decay, whether of animal or vegetable matter, and is made by every thing that burns, as the result of all combustion. Yet so constant as are its sources, a wise Providence has so ordered it, that it does not accumulate amid the open air, and is confined in deleterious quantities only under special circumstances. If a large fire of hard-coal is kindled, and the pipe touches against the back part of the chimney, gas will accumulate in the room, a portion of which is carbonic, and if at night, when the family are asleep, even from this suffocation may ensue. In our own State a case of this kind recently occurred, in which a parent was awakened just in time to save himself and all his family from death, some of whom were already suffering severely from its effects. In burning or slaking lime the same gas is given off, and where a large number of persons are confined for a long time in a close room the same effects are produced. In this class of cases the amount is not so great as to cause the forcible and persistent closure of the breathing orifice, but this foul air is gradually inhaled, the change of air in the lungs prevented by the absence of pure air, and if long continued, death must follow from the circulation of carbonic acid in the system.

Persons who descend into deep wells or pits, will be lowered without inconvenience until the mouth comes to the level of the height of the foul air, and then fall suddenly and senseless to the bottom; and no one should ever venture upon such a descent, until it has been found by trial, just before, that a candle will burn freely. The carbonic gas may be removed, by forcing in a draught of pure air, by pumping out the foul air, or by dashing water around the sides of the empty pit, or by such other chemical means as it may be convenient or necessary to apply.

But the most common cause of fatal results from carbonic acid gas, is the burning of charcoal in a furnace. This consists almost entirely of pure carbon, and in its lighting and burning, large quantities of the gas are evolved. This, in a close room, will soon accumulate, and amid the pleasure of the warmth, the victim is lulled into a fatal lethargy.

In all these cases, the danger arises from the insidiousness and suddenness of the effect, and from the soothing power of the gas in small quantities. Thus, when carbonic acid is introduced into the sleeping room, it inclines the person to sleep more readily, the mind to act with less forethought, and thus prepares the unfortunate one to forget the cause, and sleep himself into eternity. Even where the subject is awake and active, the effect is very rapid. A few months since, in my own vicinity, several were suddenly, and for a time seriously affected in this way. On the occasion of a marriage gathering, a charcoal furnace had been placed in one of the rooms having no fire-place, for the purpose of warming it, and several of the guests were conversing together in the room. All at once one of them experienced peculiar sensations, immediately two or three others were similarly affected, and without suspecting the cause, all found themselves feeling seriously ill. The origin of the trouble being discovered, curative means were immediately employed, and all relieved, some, however, only after several hours of unpleasant sickness. A single example like this serves to show how insidious is the action of this invisible destroyer, and how important that the sources of danger may be recognized and avoided.

Where a person has become thus affected by carbonic acid, the first indication is, to remove the cause, and place him where pure air can be obtained. Let him be quickly removed to another room, the windows raised, and not too many persons around him, so as to obstruct the air, or to furnish him with a new supply of the gas from their own lungs. In the case of a person let down into a well, if no answer is given to a call, let him be rapidly and carefully pulled up. If no rope has been attached, let another down, who may attempt to drag him from his death-bearing atmosphere, and who himself, if affected, may be instantly withdrawn.

If it is deemed hazardous for any one to descend, dash a pail of cold water down every two or three minutes from above, and with grappling-irons, or a bent farm-yard fork, attempt to get hold in some of the clothing, and raise the victim from his otherwise certain doom.

With the patient before us, and placed in a favorable atmosphere, all other means used must have reference to two points—the ridding of the lungs of foul air they contain, and the causing of free, full inspirations of pure air. The neckerchief should be removed, the clothing unloosed, a little cold water suddenly and repeatedly dashed upon the face and chest, hartshorn or vinegar applied to the nostrils, and they tickled with a feather, and the extremities, if cold, warmed by hot applications. If these means do not rapidly succeed, recourse must be had to artificial respiration, a minute account of which will be found under the subject of drowning. After breathing is re-established, mild stimulants, moderate warmth, and friction upon the surface is indicated.

The person will often remain weak and exhausted for several days, and tonics and care be required to overcome the shock which the system has evidently experienced.

V.—*Drowning*, is the most common form of suffocation, and one, the treatment of which will more fully explain the means to be used in all cases of obstructed respiration. There are always two prominent and leading indications in every case of suspended respiration; viz., to restore the breathing, and to use means to excite the circulation. These are so intimately connected, that the means conducive to both need to be used conjointly. The Royal Humane Society have, for many years, published a code of instruction for the recovery of those apparently drowned, but the investigations of Marshall Hall have fully demonstrated that, in some particulars, their directions were defective, and those issued by the

National Life Boat Institution, founded upon the experiments of Hall, have proven much more efficacious.
We cannot do better than here to transcribe the code
of management which they direct.

"I.—Treat the patient instantly, on the spot, in the
open air, exposing the face and chest to the breeze, except in severe weather.

"II.—To clean the throat, place the patient gently face
downward, with one wrist under the forehead, in which
position all fluids will escape by the mouth, and the
tongue itself will fall forward, leaving the entrance into
the windpipe free. Assist this operation, by wiping and
cleansing the mouth.

" If there be breathing, wait and watch ; if not, or if
it fail, then—

" III.—To excite respiration, turn the patient well and
instantly on the side, and

" IV.—Excite the nostrils with snuff, hartshorn, volatile
salts, or the throat with a feather, &c., and dash cold
water on the face previously rubbed warm.

" If there be no success, lose not a moment, but instantly begin,

" V.—To imitate respiration. Replace the patient on
the face, raising and supporting the chest well, on a
folded coat, or other article of dress;

" VI.—Turn the body very gently on the side and a
little beyond, and then briskly on the face, alternately,
repeating these measures deliberately, efficiently and
perseveringly, about fifteen times in the minute, or every
four seconds, occasionally varying the side. (By placing
the patient on the chest, its cavity is compressed by the

weight of the body, and expiration takes place. When turned on the side, the pressure is removed, and respiration occurs.)

"VII.—On each occasion that the body is replaced on the face, make uniform but efficient pressure, with brisk movement on the back, between and below the shoulder blades or bones on each side, removing the pressure immediately before turning the body on the side. (The first measure increases the expiration, the second commences inspiration.)

* * "The result is, respiration, or natural breathing; and, if not too late—life.

"VIII.—After respiration has been restored, promote the warmth of the body by the application of hot flannels, bottles or bladders of hot water, heated bricks, &c., to the pit of the stomach, the arm-pits, between the thighs, and to the soles of the feet.

"IX.—To induce circulation and warmth. During the whole time do not cease to rub the limbs upward, with firm, grasping pressure, and with energy, using handkerchiefs, flannels, &c. (By this means the blood is propelled along the veins toward the heart.)

"X.—Let the limbs be thus warmed and dried, and then clothed, the bystanders supplying the requisite garments.

" *Cautions.*—1. Send quickly for medical assistance and dry clothing.

" 2. Avoid all rough usage and turning the body on the back, i. e., flatly and permanently.

" 3. Under no circumstances hold up the body by the feet;

"4. Nor roll the body on casks;

"5. Nor rub the body with salts or spirits;

"6. Nor inject tobacco-smoke or infusion of tobacco.

"7. Avoid the continuous warm bath.

"8. Be particularly careful to prevent persons crowding around the body.

"*General Observations.*—On the restoration of life, a teaspoonful of warm water should be given, and then, if the power of swallowing have returned, small quantities of wine or brandy and warm water or coffee. The patient should be kept in bed, and a disposition to sleep encouraged.

"The treatment recommended should be persevered in for a considerable time, as it is an erroneous opinion, that persons are irrecoverable because life does not soon make its appearance, cases having been successfully treated after persevering several hours."

This we believe to be the best plan in the hands of one skilled in its application; but the following, known as the Silvester method, will be found by some more simple and more readily applicable.

"1. Position: Place the patient on his back, with the *shoulders raised* and *supported* on a folded article of dress.

"2. To maintain a free entrance of air into the windpipe. Draw forward the tongue, and keep it projecting beyond the lips. By raising the lower jaw, the teeth may be made to hold it in the proper position.

"3. To imitate the movements of respiration: Raise the patient's arms upward by the sides of his head, and then extend them gently and steadily upward and forward for a few moments. (This action, by enlarging the capacity of the chest, induces inspiration.)

" Next turn down the arms, and press them gently and firmly for a few moments against the sides of the chest. (Forced expiration is thus effected.)

" Repeat these measures alternately, deliberately and perseveringly fifteen times in a minute."

One or the other of these methods will often succeed under circumstances apparently hopeless.

In the asphyxia of new-born infants the same general plan may be adopted. The following are the directions given by Marshall Hall on this point:

" 1. Place the infant on the face.

" 2. Sprinkle the general surface briskly with cold water.

" 3. Make gentle pressure on the back; remove it, and turn the infant on the side; and again place it prone with pressure.

" 4. Rub the limbs, with gentle pressure, upward.

" 5. Repeat the sprinkling only now and then with cold and hot water (of the temperatures of 60° and 100° Fahr.) alternately.

" 6. Continue these measures, or renew them from time to time, even for hours. The embers of life may not be entirely extinct !"

8

When any portion of the intestine presses out of the cavity which was intended for it, so as to be felt as a bunch either near the groin or at the navel, the person is said to be ruptured. This was once considered quite disgraceful, for such men were not competent as soldiers, not being fitted for long marches; but since virtue and valor have come to be viewed as different things, and greatness to consist in mental and moral, rather than in physical qualities, the disgrace of the thing has vanished, and the more since it has been found that about one-ninth of all men are troubled in this way. Sometimes children are born with them, but oftener they are the result of a strain or blow, or some sudden exertion. The bowel at some point where two or three muscles meet or cross each other, by some accident finds its way between them, and protrudes so as to be covered only by the skin and the substance immediately beneath it. Most persons when they have a difficulty of this kind, soon become sensible of it by some uneasiness, or by discovering a slight enlargement; and it is a false delicacy in men or women to conceal the difficulty, as neglect is often attended with most serious conse-_quences.

They are generally easily known by their position, as already alluded to, either at the navel or near the region of the groin. A slight cough will cause the little tumor to feel more full; and when the rupture is recent, upon lying down the lump will recede. If no means are used to prevent it, they generally go on increasing in size, in some instances until almost the entire bowels leave the cavity intended for them, and occupy this new and uncomfortable position. In some cases, even of long standing, upon assuming the horizontal position, the mass will go back to its place. It is then known as reducible hernia or rupture. In other cases the gut becomes fastened to the sides of its new residence by strips of membrane which are there formed, and then it becomes irreducible; and in still other and more dangerous circumstances, the bowel thus down becomes filled, so as to press severely against the side of the ring through which it has passed, and thus prevent the circulation of the blood, and then it is affected as if a string were tied about it. If not relieved, the result in this case must be at no very distant time death; and any rupture which is not prevented from coming down, is liable at any time to assume this character. Even where no such result follows, the dragging weight thus occasioned is not only uncomfortable, but is apt to affect other organs, and interfere with the process of digeston. All these considerations render this subject important, and deserving of the attention both of the patient and physician. With proper knowledge these can almost always be prevented from becoming serious, and even when strangulated, a due sense of the danger will lead you to seek more quickly for medical aid, and thus be more likely in time to secure efficient relief.

A recent rupture is generally easily reduced by lying down upon the back, and making gentle pressure over it in the direction opposed to that in which the bowel has descended. If this do not succeed, lie down upon

the floor, and run the feet up against the side of the
wall, thus partially turning the body down side up, and
by this means favoring the return of the mass. If this
do not avail, and any pain is experienced, medical aid
should be quickly sought, and when it has been reduced
you should at once seek the means for preventing the
descent of the mass.

Most cases which have not become irreducible by adhesion, by means of bands to the sides of the sac, admit
of permanent cure, and those which do not, may be pre
vented from causing any trouble.

An instrument known as a truss, is the easy and
effective remedy, if properly made and fitted, and constantly worn, but if not, a very uncomfortable and un-
satisfactory thing. If the pad is flat, it merely covers
the hole, like a saucer over the mouth of a jug, instead
of being shaped like the small end of an egg, not only
covering, but pressing up into the cavity. By this
means, the flesh thus pressed, is united to the side of the
opening, and effectually stops up the hole. The spring
should be elastic, and fitting to the shape of the surface
on which it presses, and the angle of the pad to it such
as to cause a steady, even pressure, and the whole so
protected as not to irritate the parts. India rubber furnishes a good pad, and steel, well covered with chamois
skin, the best elastic spring. Generally, a physician
should fit it to the part, as he is best acquainted with
the kind of truss needed, and as all its value depends
upon its adaptation.

In a reducible hernia, it should always be the aim to
effect a cure, especially in those not having reached the

middle period of life. In applying the truss, the rupture should first be reduced, then the truss applied, and the patient, by standing up and coughing, can determine whether it accomplishes its first object of keeping up the intestine. For a while there will be some irritation from the pad or spring, and the part should be bathed each night, after lying down, with a little brandy, or borax water, and if chafing occurs, sugar-of-lead wash will need to be used, and a linen rag to be placed beneath the pad.

The truss should be kept on night and day, where a permanent cure is sought, and it is well to have a second one, so that, if the first is broken, the other can be immediately adjusted. By leaving it off a few hours, the work of cure, which, unseen by us, has been going on for months, may be all undone. A year or two will often effect entire restoration, and the result, even after so long a time, is worthy of the effort. If a person, having a reducible rupture, gets in trouble therefrom, he, or some one else, is usually to blame; for, even if a complete cure is not obtained by a proper truss, properly and continually worn, danger therefrom may be prevented, except in case of accident. But a rupture may be irreducible from the first, or, being left down, it may be impossible to return it, on account of swelling, or of more food or blood pressing into it than is able to return, or it may become permanently irreducible, by adhering to its sac, as already noticed.

It is very rare that a rupture from the first start is irreducible. The patient can generally, by a little pressure, and by means already alluded to return it;

but if not, let him get into a warm bath, and continue attempts to reduce it until his physician arrives. Where a rupture that has been reducible from being left unsupported by a proper truss becomes irreducible, the same methods heretofore noticed must be employed. Cold applied in the form of ice, over the mass, and an injection of flax-seed tea may be serviceable; but bleeding, the careful and skillful pressure of the surgeon, or even an operation may be required, and no time should be lost in securing aid. Ruptures are especially liable to be troublesome after eating very heartily, especially of vegetable food, and beans, corn, cabbage and turnips need most to be avoided.

Where a rupture has become permanently irreducible by means of adhesion to the sides of its sac, the patient may live for years with it in this condition, but is liable at any time to serious trouble from its becoming strangulated. If small, a pad should still be worn to prevent enlargement, or if protruding considerably, a bandage or suspender should be worn so as to prevent dragging and still further protrusion. Even these cases before middle life, admit of some relief, and various means besides those already noticed have been used both for the relief and radical cure of reducible and irreducible ruptures; but as these need always the personal supervision of the physician, extended notice of them now is not required. The symptoms of a strangulated rupture are as follows:

1st. If before reducible, it is now irreducible.

2d. There is a feeling of unusual constriction about the part, as if a cord had been drawn tightly about it, and a general feeling of uneasiness. Hiccup or raising of wind, paleness and some prostration soon occur. The pain is more severe if partially strangulated, than when suddenly and severely. Other symptoms, such as vomiting, rapid small pulse, swelling of the bowels, cold sweats and irregular breathing soon supervene, and too often, all efforts at relief fail, and even a skillful operation does not relieve. This, however, if resorted to in time, will often snatch the patient from the jaws of death, and there must be no delay in procuring surgical assistance. By the peril of such a position, and the accessible possibility of avoiding it by the proper and persistent use of the truss, we would again entreat all those who have reducible ruptures, never to run this risk; and those who have irreducible ones, constantly to be on the guard to prevent their becoming strangulated. In the case of children, parents should be particularly anxious to attempt cure early in life, as chances are much in its favor. A well-adjusted truss with them, is almost certain to effect a complete relief; and when the rupture is at the navel, a small half hickory nut sewed into a bandage, the oval side applied over the point, and the bandage fitting nicely about the body, will generally prevent further trouble. Prevention is said to be better than cure, but cure itself is far better than the risk of a constricted intestine. Let no one, then, neglect for themselves or those committed to their care, that relief which cures the rupture or prevents serious consequences therefrom.

INCONTINENCE OF URINE.

This may depend either upon irritability of the bladder, or upon paralysis. The most frequent and troublesome form in which it occurs is with small children, and with them it almost always results either from general weakness, or an irritable state of the ends of the water tubes as they open upon the surface of the bladder. After the difficulty has existed for some time, it is often perpetuated as a mere habit, much to the annoyance of the child and of its bedroom attendants. Often children are severely punished for this failing, but while it is well to bring their mental effort to prevent and to bear upon the case, yet punishment for disease is very seldom required.

The child should be allowed no drink at the evening meal; should be taught not to lie upon the back, but upon the side, or still better, over upon the belly. The bladder should be emptied the last thing before retiring, and the child taken up again before the usual time of discharge, as this is before midnight. If these means, persevered in, do not succeed, spread a fly blister, the size of a dollar, sprinkle it over with camphor, and apply it directly over the region of the bladder until it draws a slight blister. In addition to these means, it is often well to use, as an internal remedy, a little of the muriated tincture of iron, but this, like most internal remedies, should be taken under the direct advice of a physician. None of these means should be left untried, for children often suffer much in their health, as well as in their comfort, from the annoying habit.

INJECTIONS.

Injections or clysters are often a valuable means of relieving the intestines. Where there is an accumulation in the lower bowel or rectum, it is more readily reached by these than by medicines given by the mouth. Sometimes there is a want of tone or strength in the lower part of the intestinal canal, and an occasional injection of cold water will often relieve constipation.

An injection-pipe is not expensive, is easily managed, and should form a part of the ware of every family. The most common one in use is a metal tube, having a piston which sucks up the fluid, and discharges precisely on the principle of a child's pop-gun. You may fill it in either of two ways: Having pushed the piston down, hold the end of the tube in the basin of liquid, and then by drawing up the piston, you fill it. Where it will not suck up the water, it is generally owing to a want of tightness of the box, and a little oil, or the wrapping of a little lampwick smoothly around the piston, will cor_ rect the difficulty.

Another method is entirely to withdraw the piston, place your finger over the lower opening of the tube, and pour in from the top the liquid to be used until the tube is nearly full. Then put in the piston and screw on the top, and grease the small part of the tube. The small part of the pipe should then be introduced into the intestine, inclining back and a little to your right. Then holding it firmly, let the piston be pushed steadily down, and thus the fluid is thrown with more or less

8*

force into the intestine. More expensive and convenient
contrivances may be had if desired, but all operate upon
the same principle. If the injected fluid does not remain, but is rejected at once, let more be very slowly
thrown up. For an adult, from a pint to a quart at a
time is not too much, and in urgent cases, quart after
quart will be of service. As to the ingredients used,
these are various according to the design to be attained.

Pure warm water is a good injection. If you wish a
cathartic effect, add of castor oil, sweet oil, spirits of
turpentine or molasses, three times as much as you
would give by the mouth.

If the bowels are very inactive, a stimulating injection may be needed. For this purpose, add to warm
water, or to either of the others, a tablespoonful of salt,
or a piece of soap as large as a walnut, dissolved. If
there is much collection of wind, which often occasions
pain, a teaspoonful of laudanum for an adult, or of soda,
or of essence of peppermint, or of tincture of assafœtida,
or all combined, with the water, will be of service. If
much pain, laudanum may be freely added in treble the
quantity of a usual dose. Where there is great debility,
and the design is to nourish, milk, wine, and egg make
the best clyster. Where no other injecting apparatus
is at hand, or you have not the money to spare to buy
one, a pig's bladder filled with the liquid to be used, and
an elder tube tied in it, will answer every purpose.
Then introduce the tube into the rectum, and squeezing
upon the bladder, the liquid is forced out. Any one
who has the least common sense, can with these directions administer an injection to himself or others; and

this valuable means of relieving pain and constipation should not be lost sight of. With quite small children, they are often much preferable to the administration of medicine. Pain and costiveness not unfrequently are caused by an accumulation of hard fœcal matter in the lower bowel, which can most readily, and sometimes only be removed by the softening and contractile action which an injection occasions, and they are not unfrequently the means, when persevered in, of returning the intestine to that healthy condition which prompts it each day to rid itself of its contents. Injections may easily be self-administered, by filling the tube and pressing the piston against a low stool, or by the patent syringe, which enables you to have most of the apparatus in front. False notions and prejudices exist against these most excellent curative agencies, and they often form a ready and effectual means of relief.

BROKEN AND DISPLACED BONES.

Where a bone is broken, or a joint dislocated, there is always considerable pain; and if the person has to be carried any distance, it is very important that it be done in such a way as to secure the greatest possible comfort. When he has need to be conveyed far, an easy spring carriage should be procured, a soft bed placed in the bottom of it, and the patient laid thereupon, the limbs being propped and supported by straw or pads, in such a manner as will give the greatest ease. Haste and

excitement are to be avoided, or, at least, comfort not to
be sacrificed thereto. If the home is near at hand, or
the injury one occasioning great pain, the patient can be
borne more comfortably upon a fixture known as a hur-
dle. This may be made extempore, by taking a common
cot and fastening a strip at each end of it so that it will
not fold, and then conveying the sufferer thereupon. A
long shutter or plank will answer a similar purpose. If
the patient is sensible, he is usually the best judge of
the most comfortable position, and his ease should be
consulted. Often much injury is done by a careless
conveyance. If the bone has not thrust through the
skin, it is very desirable so to carry the person that it
should not; and careful support by pads, or the use of
two or three strips of pasteboard or cigar box, fastened
so as extend above and below the fracture, will often be
of great service. If the bone is thrust out, and there
is dirt or dust on the end of it, this should be brushed
off, and even a couple of the bystanders, if a physician is
not at hand, by a little pulling in a straight line, above
and below the point, may so far reduce it as to cause it to
feel less uncomfortable. The pain from a broken bone at
first is very little, except from displacement of the parts,
and stiff pasteboard or cigar box wrapped about and fastened on, will serve to keep them in place. In dislocation, that is, where the bone is out of joint, it should be
supported in a sling in such a way as will be least painful to the patient. The bed, in either of these cases, on
which the person is to be placed, should, if circumstances will permit, be prepared at once so that he may not
need to be moved therefrom. A straw bed or mat-

itress is to be preferred. Much suffering, and often consequent displacement, results from frequent movings; and careful preparation beforehand, and careful removal of the garments from the wounded part, save much unnecessary pain. If there is much faintness or exhaustion, brandy may be given freely every few minutes, but this must be discontinued as soon as there is reaction. A broken or displaced bone is always best to be put in place at once; but in case of fracture, for the first few days it should not be fastened too firmly, since swelling and inflammation are apt to ensue. As a fracture or dislocation has to do with the stay-work of the frame, the friends of the person concerned should early seek surgical aid, as a neglect of a few hours or days may permit alterations of structure to take place which no after means can fully overcome.

SUN STROKE.

Sun stroke, or *coup de soliel*, is a disease, the precise nature of which has long been a disputed question among medical men, but its results are so sudden and fatal, that it is, at least, evident, that it touches the very main springs of human life.

Sometimes it is accompanied with fullness about the head, and other symptoms of cerebral congestion, but, usually, the prominent effect is a general prostration. The countenance is pale, the heart enfeebled in its action, and the blow seems aimed at the nervous system, producing a state much akin to that we find in the last stages of nervous fever.

Persons suffering from headache, diarrhea, or from some other cause not feeling as well as usual, are most liable to fall victims to its power. Large draughts of cold water, full meat diet, and alcoholic stimuli, especially predispose to the attack. Sir Charles Napier, while serving in Sindh, was himself severely seized, and in finishing a letter, a part of which he had written beforehand, thus speaks of his case. "I had hardly written the above sentence, ten days ago, when I was tumbled over by the heat; forty-three others were struck, all Europeans, and all died within three hours, except myself! I do not drink!—that is the secret. The sun had no ally in the liquor amongst my brains." The disease is especially prevalent in India, and it has been plainly shown, that the avoidance of spirit rations, regularity in eating, and the use of a thin flannel suit next to the skin, have much to do with its prevention. The head-covering should be light, not fitting closely to the head, and so open as not to interfere with due ventilation. Fishermen, in hot days, not unfrequently put moist sea-weed in their hats, but any dry, non-conducting surface will serve as well. We see no reason why head-refrigerators are not as practicable as some of those for other purposes.

There are, sometimes, premonitory symptoms, which, if noted in time, may lead to such measures as will ward off the attack.

Headache, dizziness, a general uncomfortable sensation, an oppression of breathing, and a "feeling the heat" more than usual, should lead to suspicion and protection. When a person, apparently overcome by heat, suddenly falls senseless, at once let cold water be applied to the head, the neck-cloth be loosened, the breast exposed, and the water be sprinkled freely thereupon. Move him, in a recumbent position, to a shady and airy place. Offer the patient a teaspoonful of cold water, and if you find he can swallow, administer a half-tablespoonful of brandy, with a teaspoonful of essence of peppermint, and a half-dozen drops of laudanum in it. Mustard, dissolved in water, or spirits of turpentine, should be rubbed along the spine, warmth applied to the four extremities, and general rubbing of the surface, with the hands or dry flannel, be persevered in. Ammonia or hartshorn should be applied to the nostrils, and a weak solution of it may be administered internally, instead of other stimuli, the strength of which, first test by tasting of it yourself. Where the patient cannot swallow, a pint injection of milk punch, to which a teaspoonful of turpentine and peppermint have been added, are most proper. Even when the case seems hopeless, these measures should not be omitted, and artificial respiration should be resorted to, as directed for drowning and other cases of suspended animation. Recoveries are recorded as having taken place under the most unpromising circumstances.

Sometimes the attack is not so sudden, and the person will feel himself to be overcome with the heat, without the loss of consciousness. Here at once cold water should be even more freely used, externally, the feet be placed in a warm bath, and general friction to the surface employed. Coffee and ginger, or cayenne pepper tea will be sufficiently stimulating. While any means are being employed, medical aid should be sought, for there are variations of treatment which particular symptoms designate, and leeching, cathartics and opiates are not unfrequently required. Those who recover should be very careful to avoid subsequent exposure to heat, especially when direct, or reflected from hot surfaces in a crowded locality.

EFFECTS OF FALLS.

Adults, by accident, and children by their venturesome propensities, not unfrequently are subjected to severe falls, and these, when not seeming very serious at the time, afterward prove the cause of most urgent symptoms. Where the injury done is internal and imperceptible, the part most liable to be affected is the head. The portion bruised should be freely and fre quently bathed with cold water, or spirits and water, and the person for a time kept quiet in the recumbent posture. All noise and excitement should be avoided, and the light partially excluded from the room. If there is faintness, a little brandy may be given, but, if possible, external applications and frictions should be relied upon in its stead. Undisturbed rest will often prevent difficulty, even where the jarring has been considerable.

If *vomiting* supervenes, it is an almost certain symptom that there has been concussion of the brain, more or less severe, and it will be well to seek the advice of the physician. Where there is compression of the brain, as indicated by stupor and unconsciousness, cold to the head, warmth to the extremities, a mustard draught to the back part of the neck, and friction of the surface will be judicious until medical aid shall arrive. For a long time after a severe fall, it is important to avoid excitement of any kind. Study should be relaxed, the temper unruffled, exposure to the heat avoided, the child restrained from overplay, the bowels regulated, and special attention given to keeping the skin in a healthy condition. Many a case of dropsy of the brain is developed by a neglect of these precautions.

APOPLEXY.

As this is often very sudden in its attack, it is well for all to know the means immediately to be employed. When a person without either spasmodic action or faintness suddenly falls senseless, it is generally owing either to a congestion of the brain, a rush of blood to the head, as it is called, or to an actual rupture of a small vessel in the brain pouring out blood, and by pressure causing him to lose the power of thought or motion. When a friend is thus seized, at once loosen the neck handkerchief, or any article of apparel that is tight about him, and dash cold water in the face. A small pitcher of cold water poured upon the head from a height will be of benefit, but do not continue it long, especially if there is less color in the face than usual. Then apply warmth to the feet, and a mustard plaster to the back of the neck, and the breast. Rubbing the limbs, and hot applications thereto, are also of benefit, as these have a tendency to keep up, and make more active the circulation of the blood in the outside vessels. All these means may be resorted to at once, and often will much moderate the severity of the attack. The great point is to keep the head cool, and prevent the blood from settling in the larger vessels, causing congestion. The person should be laid in an easy horizontal position, with the head slightly raised. Bleeding from the arm and active purging, will usually next be necessary; but this must be left for the judgment of the physician, who should be summoned in haste.

CROUP.

As apoplexy is peculiarly a disease of age, so is croup one of the most formidable of childhood. Sudden and insidious in its approach, and hasty in its often fatal termination—not unfrequently it is seen by the physician too late for any reasonable hope of relief. True croup may be confounded with what is called false croup, and with mere hoarseness. Where there is not difficulty of breathing, no excited drawing of the breath, and merely a thickness of articulation, the case is one of cold, affecting merely the vocal chords, and easily distinguished from croup. The distinctions between false and true croup are not so easily marked out to the unprofessional person. False croup is dependent upon a perverted reflex action of the nerves, causing spasm of the muscles of the throat. The white leathery substance known as false membrane, is not thrown out in this. Croup is undoubtedly generally attributable to sudden changes of atmosphere. A child passing from a dry hot room into the cold air, or from a dry to a damp position, or cooling off suddenly when in perspiration, will frequently contract the disease; in fact, the same circumstances favorable to catching cold in an adult, will give rise to croup in a child.

When your little boy or girl has a hoarse cough, with difficulty of breathing, even these first symptoms are worthy of notice, and a foot or hip bath, with a good bowl of boneset or catnip tea, will not be amiss. If there is a ringing cough, a panting for breath, a slight wheezing, and especially a prolonged and difficult inspiration—that is, the greatest distress in drawing in, instead of breathing out the air—you may well feel some uneasiness, and an emetic will be on the safe side.

These are the only class of cases in which antimonial wine may be administered before medical advice can be obtained; and if this is not at hand, syrup of ipecac may be used every fifteen minutes until vomiting is produced. For doses, see chapter on that subject. At the same time, mustard drafts may be applied to the feet, the spine rubbed with spirits of turpentine, diluted with melted lard or sweet oil, and a rag, frequently wet with cold water, be kept just over the Adam's apple, as it is called, or a little below it. In the onset of the disease, these are the chief means to be used.

If the bowels are constipated, a mild cathartic will be of service, and if the symptoms increase in severity, bleeding may be necessary. When there is much pain in breathing, hot applications may be substituted for cold, or they may be used alternately.

These are always cases requiring the greatest medical skill, cases in which the doctor himself too often feels the weakness of human aid, and, while no time should be lost in applying the remedies we have noticed, medical assistance should be speedily procured. When true croup passes to its second stage, the chance of recovery is exceedingly limited.

Every family should keep in the house either boneset, antimonial wine, carefully labeled and put away, or syrup or powder of ipecac, so as to be ready for the disease in an emergency.

SUMMER COMPLAINTS.

During the hot season of the year, disorders of the bowels are among the most common affections, and are often so sudden and prostrating, as to require attention even before the presence of the physician can be secured. These may arise from a variety of causes, such as sudden changes of temperature, errors of diet, sourness of stomach, derangement of the functions of the liver, and, in children, from teething, and various other causes.

Where a person is seized with diarrhea, accompanied with pain and vomiting, whatever may be the cause, the indication is, to control these, and, in doing this, sedatives, astringents, and counter-irritants, are among the remedies indicated. For an adult, a good mustard plaster over the bowels, and a teaspoonful of the following mixture each half hour until relief is obtained, will usually be found of service.

Tincture of Catechu,
Aromatic Syrup of Rhubarb,
Common Soda,
Essence of Peppermint,
Laudanum,

Of each one teaspoonful.

If the mixture is not at hand, the ingredients that are, may be used in the same proportions.

Where there is simple diarrhea, without pain, a pure astringent and stimulant is all at the first needed, and a teaspoonful each of the catechu and rhubarb will generally suffice. Where there is severe pain, a full dose of some opiate, as laudanum or paregoric, is always proper.

The very fact, however, that diarrhea and sickness are dependent upon a variety of causes, shows the importance, if the symptoms do not soon abate, of seeking more intelligent aid, as cholera or dysentery may ensue.

Beside these general causes of derangement of the bowels, there are those which apply only to children. Sudden changes of temperature, and especially a high degree of heat, affect them more readily than adults, and the process of teething often gives rise to a very troublesome irritation of the stomach and intestines.

The second summer is often the dread of the fond mother, and large numbers of children fall a prey to cholera morbus, dysentery, cholera infantum, marasmus, or some other variety of bowel affection. It cannot be denied that the child is subject to many real sources of disease, either inherited, or dependent upon unavoidable causes, but we nevertheless believe, that much of the trouble is artificial, resulting from errors of diet and management.

Children who are weaned at about the age of one year, and afterward fed on proper food, are least of all subject to these affections. Those who both feed and nurse are exposed to the results of evil influences upon their own systems and that of the mother, while those placed on improper diet are sure to suffer from the change. The great point, both in the prevention and cure of these infantile summer complaints, is, so to manage the diet and patient as to place it in the most favorable circumstances for health, and while the judicious care of the doctor may be needed, he, most of all others, would desire to have you impressed with the fact, that his medicines will prove utterly valueless unless the diet and regimen are properly regulated. The inner garment of the child should be a thin woolen flannel, and a small bandage about the bowels, moderately tight, is often of service. The sleeping-room must be well aired in the daytime, and the place in which the child is kept during the day must be such that perfect ventilation can be secured. The striking contrast between the fatality of these diseases in city and country, exhibits the influence which these outside circumstances have upon them. Green vegetables should never be allowed to young children during the heat of summer. New potatoes, which cut like hard soap, cucumbers, beets, peas and beans are very common exciting causes of this disease. Solid, raw cherries, green currants and gooseberries are among the most objectionable of fruits. Milk, a little well-cooked meat once a day, mutton broth, stale wheat bread, mush that has been boiled over half an hour, and allowed to get cold, mealy potatoes, rice, good ripe fruits, butter, corn-starch and tapioca, and plain sugar cake, eaten regularly, form a sufficient variety without giving them whenever they may desire them, pies, sweetmeats, green corn, peanuts, rich cake, egg-plant, fried clams, raw onions, or any of the other delicacies of life.

Stomach arguments are, of all, the most difficult for the physician to enforce, and every now and then some proud parent will boast of a little boy who has passed the summer eating green apples and butter-milk, and relishing cabbage and turnips, with his other food, without immediate and perceptible injury therefrom. Such cases are often quoted as triumphant proofs that various articles are not hurtful, and just such reasoning is daily relied upon by many a glutton, opium eater and inebriate. Just as well might we claim that there had been no carnage on the plains of Italy, because multitudes of those engaged still survived in health and strength. In order to get at the facts, we must count the slain ; nay, more, in the bulletins of disease we must notice those cases in which, though no outward wound is seen, yet an inward effect has been produced which will, in time, bear a bitter fruit.

Upon the judicious management of the mother, even where the physician is aiding, turns, as on a pivot, the life of the child, and never more so than in regulating the dress, diet, air, sleep, and general care in summer diseases. As to the remedies to be used, the aromatic syrup of rhubarb, one tablespoonful,

> Tincture of catechu, ⎫
> Paregoric,   ⎬ of each a teaspoonful,
> Prepared chalk,   ⎭

mixed and shaken, and a teaspoonful given every three hours to a child of three years, if the evacuations are as frequent as that, forms the best home remedy; but so different are causes and accompanying symptoms, that reliance for any length of time upon any single prescription is not as safe as the opinion of the skilled practitioner.

COLIC.

Sometimes, instead of a free evacuation of the bowels, there is intense griping pain, vomiting, and general uneasiness, accompanied with constipation. This may arise from bilious disorder, from the accumulation of hard fœcal matter in the lower intestine, from indigestible food, or, as in painter's colic, from the introduction of lead slowly into the system. In all these cases the patient or his friends are very likely to conclude that the bowels must be moved, and to feel that they at least are equal to the treatment of the case thus far. Accordingly, salts, oil, and various other medicines are used for the purpose, and often only with the effect to increase the vomiting, and add new power to the pain. That the indication is to secure free evacuation of the bowels is undoubtedly correct, but in these cases cathartics are not the only remedies. The pain is dependent upon a spasmodic action of the muscular coat of the intestines, and while this lasts, often the most powerful operatives will be unavailing. Opiates must be administered, or bleeding, or the hot bath resorted to, to allay the internal spasm, or else the cathartic cannot act. It is well enough, it is true, to administer a reasonable quantity at first, that it may be in readiness at once to do its work when the pain is overcome; but far more important is it by the use of mustard externally, and of laudanum internally, first to overcome this violent action. Again, often the real cause of trouble is an accumulation of solid matter in the rectum or lower bowels, and at the start, injections of warm water by the pint can do no harm. This, with a mild cathartic, as oil and a mixture of twenty-five drops of laudanum and peppermint each, and a half teaspoonful of soda in water, given every hour, will often afford relief. See to it, then, that if you must do something before aid arrives, that what you do is sensible, so that the medical attendant will not find himself called upon to treat a worse case because of the means already used. The most violent cases of bilious colic I have ever witnessed, have been those greatly aggravated by promiscuous doses of cathartic medicine.

9

To prevent the Lead Colic, those who have occasion to paint, or in other ways work in lead, should wear an outer garment of close linen, should wash the face and hands, and rinse the mouth always before eating, and should, by frequent washing of the whole body, secure cleanliness and the removal of all particles of lead. Those who work constantly among lead, sometimes have a respirator so fixed that they shall not inhale the lead dust. Sulphuric acid possesses much power in preventing the effects of lead, and in the form of a so-called lemonade, or mixed with other drinks of the workmen, is used in many manufactories. Some persons are more easily affected than others by the lead poison, and cases are on record where even the inmates of a sitting room have suffered from their temporary exposure to the influence. Fifteen drops of aromatic sulphuric acid in a wine glass of water, taken each day, aids in the prevention of the disease.

PROTRUDING BOWEL.

Sometimes, from the relaxation of diarrhea or from some other cause, the lower portion of the bowel will suddenly show itself externally. As it is generally said, "the bowel comes down," and this, to inexperienced persons, often occasions much fright. A child who seems disposed to this difficulty, is, from the very irritation inclined to sit and strain, but this it should not be allowed to do. If there should be constipation, a mild cathartic, as magnesia or sulphur, may be given; but if not, there is no need of so much effort.

Should the bowel appear, the child should be placed over the lap, with the hips a little raised, and the cleft being pressed open, then with a cold wet cloth over the end of the finger, or of two or three of the fingers brought together in the shape of a cone, the part may be carefully pushed up. There is but little danger of doing harm, by any sensible pressure. Where this does not avail, the application of a little ice for a moment, or the washing the part with cold white-oak bark tea, or even with common strong tea, will tend to pucker and reduce it. After the intestine is returned, it is well to wash the part after each evacuation, with white-oak bark tea, or with a solution of alum or tannin, strong enough for one to perceive it by the taste. If there is very much swelling, the use of warm water and cold, first one and then the other for a minute or two, with pressure and slight oiling, may accomplish the object. If these means fail, it is best not to make any more severe efforts, but to seek greater experience and skill.

HEMORRHOIDS OR PILES.

These are distinguished as external and internal. External piles are those situated just on the margin of the anus, outside the sphincter or round muscle, and covered chiefly by the natural skin. They are called blind, because they seldom bleed—not because they cannot be seen.

Internal piles are within the sphincter muscle, not generally perceptible, and very apt to bleed.

External piles are blue or venous in their character, and if snipped off no great amount of bleeding occurs.

Internal piles used to be regarded as enlarged veins, but now are believed to be folds of thickened mucous membrane. The hemorrhage from them is arterial, and they are by far the most common and troublesome kind. The first indication in respect to piles, is to avoid constipation; and for information upon that point we refer you back to the article on that subject. Constipation is often the cause, and after they have been produced it helps to perpetuate them. The constitutional treatment of piles is the same as that of constipation. Locally, however, additional means may be used. Great care should be taken in cleansing and drying the part after a passage. Rough materials should not be used against the anus, and printer's ink upon harsh paper helps to keep up the irritation. If the pile has protruded, it should be greased with some astringent or soothing ointment, or bathed with white-oak bark tea, and pressed to its place. If the pile is completely external, it is best not to be troubled with it long, but to go directly to the surgeon and have it opened and cured. In internal or bleeding piles there is often more trouble. Being situated directly upon the sphincter or round muscle, every time there is a passage, especially if at all constipated, the membrane is stretched as the muscle opens, and bleeding ensues. Before a passage, let an injection of luke-warm water be used, and then afterward a little of the following ointment applied while the bowel is empty. It may be pushed up to the part without much difficulty, and will afford relief.

Powder of nutgalls, } Of each half a teaspoonful.
Powdered opium,

Powdered sugar of lead, one quarter of a teaspoonful.

Melted fresh butter or lard, one tablespoonful.

Mix, and when cool apply from time to time about the amount of a thimbleful. Where there is much constipation, thirty grains of the extract of hyoscyamus should be substituted for the opium. If this does not avail, the only alternative is to have them removed by ligature.

When piles are in an inflamed state, causing constant pain and uneasiness, even when quiet, a course of treatment must be adopted suited to this particular condition. Leeches to the anus, a dose of castor oil, linseed oil and spirits of turpentine, a half tablespoonful each, or warm fomentations and poultices to the anal region, will be of service.

SKIN DISEASES.

Under this head are included a large number of affections which, though differing much from each other, are nevertheless susceptible of classification. Some of them are so mild as scarcely to be worthy of notice, others force themselves upon our attention as troublesome, rather than dangerous, while still others are among the

most formidable and fatal forms of disease. It shall be
our aim only to name the most important, to designate
the symptoms which betoken their approach, so that aid
may be sought in time, and to notice briefly such points
of treatment as may be appreciated by the patient or
his friends, in the absence of the medical attendant.
We have already had occasion to notice the important
relationship the skin bears to other portions of the body,
and how quickly its diseases manifest themselves in general sympathy throughout the system. This is but
made more evident, as we come to study these dermoid
affections, and for this reason their management is often
a subject of vital importance. Commencing with those
most formidable and frequent, we shall refer to the
prominent characteristics of each.

VARIOLA, OR SMALL POX.

This, once the dreaded scourge of nations, before
which armies fled, and the ravages of which brought
terror into crowded cities and rural hamlets, has, by the
efficiency of inoculation and vaccination, become comparatively a rare disease. Even yet, however, it is not
a very popular affection, and both from its serious
nature, and the effect which it produces where life is
preserved, it is worthy of careful notice. It is eminently a contagious disease, and usually commences
from twelve to fourteen days after exposure. The first
symptoms are, severe aching pains in the limbs, but
especially in the lower part of the back, and the region
of the loins, flashes of heat and cold, vomiting, headache, and general constitutional disturbance. In the

onset, convulsions in children, and delirium in adults, are not uncommon. The eruption appears first on the face, about the third day, and spreads in the course of a few hours over the rest of the body. At its first appearance it is more or less cone-like; a close examination will detect a semi-transparent fluid within the raised surface, and by the third day a depression is observable in the centre. The pointed pimple filling with a secretion of semi-fluid matter, and then becoming pitted in the centre, is the most distinctive mark of small pox.

In Varioloid, which is small pox modified by vaccination, we have the same kind of eruption, but it may be distinguished by its greater mildness.

In bad cases of small pox, the throat often becomes thickly studded with pustules, and much of the danger often arises from severe inflammation here. Notwithstanding that people generally are better pleased with eruptions, when, as they say, they "are well broken out," yet in small pox the severity of the disease is greater in the fullest eruption.

Suppuration, or discharge of matter, usually begins about the sixth day after breaking out, at which time the features are enormously distorted by the swelling, and the face at length blackened by the drying scabs. Where the pustules are not run together, and the progress of the disease is regular, small pox is not a very fatal disease. It seldom, at its onset, needs active treatment. With the first symptoms, the bowels should be opened by a mild purge, as salts or cream of tartar, and coolness and cleanliness stand highest on the list of specifics. When the eruption is slow in appearing, and the patient suffering much, warm boneset or catnep tea are of service, but these must not be continued long. When the throat is painful, a weak gargle of laudanum and sugar of lead, will be of service. The condition of the eyes should from time to time be noticed, as this disease sometimes causes blindness, and local treatment may be needed for them. During the progress of the disease the diet should be low, and the bowels kept regular, but not free; but as soon as suppuration has taken place, or if there is much debility at this stage; a generous diet should be allowed. The separation or peeling off of the scarf-skin, should be promoted by an occasional warm bath. To prevent excessive pitting, the parts should not be exposed to sudden blasts of air, the pimples on the face should be opened with a fine needle when they fill with matter, and a little sweet oil or almond oil should be smeared twice a day over each pustule.

As the physician usually, on account of the anxiety of his other patients, lest he shall convey the disease, is not apt to see small pox cases as often as others, unless they are positively dangerous, it is well for the patient to be acquainted with these points, and also with the fact that such cases do not generally need active treatment, or very frequent attendance.

Chicken Pox, or Varicella, is a disease requiring no other treatment but keeping the patient quiet in a comfortable room, spare diet, and a Seidlitz powder, if the bowels are constipated. Its chief interest is the liability of confounding it with small pox or varioloid. The points of distinction are as follows:

· I.—The symptoms are much milder than those of either.

II.—The eruption first appears on the body, instead of on the face.

III.—It is more watery, and transparent, less regular, and the disease passes rapidly through its different stages to recovery, so as often to be over in six days.

MEASLES.

This is another of the contagious skin diseases. The first symptoms of its approach generally appear in from ten to twelve days after exposure. In its onset, as in small pox, there is pain of the limbs, and general disturbance of the system, but there are added thereto the symptoms of a severe cold in the head. Sneezing, watery-eyes and nose, a dry hacking cough, and a whitish-coated tongue, almost always accompany the eruption. This latter appears about the third or fourth day, commencing upon the face, and spreading over the rest of the body. The eruption is more irregular than that of small pox or scarlet fever. There will often be small patches of skin here and there which have no eruption at all, while others are completely covered. Red points are generally found in the region of the palate, before, or as soon as there is the least affection of the skin. On the second or third day after appearance it attains its height, and abates by the fifth or sixth, on parts where it first appeared.

The chief danger from measles is an imperfect eruption—its receding too soon after appearing—inflammation of the lungs, bronchitis, and diarrhea. As in other skin diseases, there may, also, be subsequent affection of the eyes or ears.

9*

The early treatment of measles is usually simple. Warm boneset tea, to bring out the eruption, and to act as an emetic, is of good service. If the eruption is tardy, and there is much pain in the head, a warm bath will be of much value. The bowels need to be moved only by a mild cathartic, and not even that, if there is no constipation. There is generally a tendency to diarrhea in the latter stages of the disease.

SCARLATINA.

Scarlet fever and scarlet rash are one and the same disease, and these different terms only designate the extent of its severity. Where the throat is not much affected, and the disease is light, it is known as scarlet rash, or simple scarlet fever; if there is much difficulty of swallowing, it is regarded as scarlet fever more fully developed; while, where the symptoms are intense, and the throat affection most severe, it is known as malignant scarlatina.

The first is often so mild, that the "tincture of time" and a little good care are alone required; the last so severe, as to baffle and dishearten the most skilled and experienced observers.

The onset of scarlet fever, even when the disease itself is mild, is generally more severe than measles. It appears sooner after exposure to the contagion than any other acute skin disease. Symptoms often appear as soon as the third or fourth day, and seldom later than the sixth or seventh. The eruption generally takes place within twenty-four hours after the first symptoms of sickness. At first there is great uneasiness, thirst, intense heat of skin, headache, and redness of the tip and edges of the tongue. The eruption spreads rapidly over the body, in the form of minute spots or points, on a bright lobster-red surface. At times, especially toward evening, the redness of a deep tinge will diffuse itself over every part. The heat of the body rises higher in scarlet fever than in any other disease, as may be shown by a small thermometer placed in the armpit. As the disease is rapid in its approach, it is also short in its duration, generally attaining its height by the fourth day of eruption.

In the severer form of scarlet fever, the eruption is more tardy in its appearance, more irregular, the general symptoms more severe, and, especially, difficulty about the throat more decided.

Scarlatina maligna is but an intensity of the same symptoms, accompanied with a greater degree of depression, and death generally takes place by the fourth day.

The early treatment of scarlet fever depends entirely upon its severity. In the mild form, unaccompanied by sore throat, protection from cold wind, but a cool temperature, drinks of lukewarm, slippery elm or flax-seed tea, or cold water, and a mild cathartic, as magnesia or oil, are all that are required. Bathing with cold water, so long as it does not chill the patient, is always a safe remedy in scarlet fever, where there is great heat of surface and dryness of skin. These often prevent the appearance of the eruption, and frequent sponging with lukewarm or cool water under these circumstances will be highly advantageous. The diet should be light, and

the bowels not disturbed much by medicines, for in
almost all of these acute skin diseases, there is a corresponding irritation of the mucous membrane of the
stomach and alimentary canal, which often shows itself
by diarrhea in the progress of the disease, and let them
rest is the rule, unless there is constipation.

The people of the present day give a great deal more
medicine than good doctors. When the throat is first
becoming affected materially, it is well to apply a nap-
kin, wrung out in cold water, and frequently changed;
but if this is uncomfortable to the patient, and there is
severe pain and swelling, it should be replaced by a
hot flax-seed poultice. A teaspoonful of vinegar, one of
molasses, and a half a one of salt, with enough water
added to reduce its sharpness, forms an excellent gargle;
and if the child is too small to use it in this way, it may
be allowed to swallow part of a teaspoonful quite frequently. Where the throat is ulcerated and the breath
foul, warm sage tea, to which cayenne pepper has been
added, in proportion of a five-cent piece full to one half
pint, should be substituted for the former mixture, and
used in the same way. It will occasion a burning sensation for a time, but this will help the condition of the
throat. There is no disease in which the careful judgment of a skillful physician, and the watchful care of an
attentive nurse, are more required than in the severe
forms of this fever, and cases which give time for treatment, are not to be regarded as hopeless.

After all the acute symptoms of scarlet fever have
subsided, and the patient seems doing well, general
dropsical effusion will sometimes supervene, and this

even in mild cases. Frequent doses of cream of tartar or spirits of nitre, with a warm bath each night, will usually relieve this ; but in many instances it will require renewed and efficient medical treatment. After any of these acute skin diseases, too great care cannot be taken to clothe the child comfortably, and to prevent exposure to dews, rains or dampness. Belladonna has acquired considerable reputation as a preventive in scarlet fever, and although many consider it worthless, my own experience has led me to value it, if given early and perseveringly from the time of exposure. As it is a remedy of some power, it had better be prescribed under medical direction.

ERYSIPELAS.

This is a non-contagious disease, and yet it is not best to wipe your face with the towel used by a person having it, or to rub a wound you have upon your finger against any portion of the inflamed skin.

In common view there are many eruptions known as erysipelas which are entirely different therefrom. True erysipelas, whether local or general, is always accompanied with swelling, and some pain or soreness, and this, of itself, will distinguish it from redness of the surface, pimples, &c., which are sometimes surmounted by this large and frightful name.

It is usually preceded by a general feeling of uneasiness, nausea, a coated tongue, fever, constipation, and a disordered stomach.

The eruption is of a deep red color, accompanied

with a sharp, pricking sensation, often sore to the touch, and generally vesicles, or small bladders of water, form upon the surface. The disease is sometimes confined to a part of the body, sometimes rapidly spreads over the surface, sometimes so far as the local difficulty is concerned, only skin-deep; but in the rare and more severe forms, it involves the tissue beneath, and goes on to suppuration or gangrene, accompanied with typhoid symptoms. Of the simple form of erysipelas, that of the face and scalp is the most common, and may become dangerous from its being complicated with inflammation of the brain, or from the prostration arising from its being diffused over a large surface. When erysipelas is mild in its character, the bowels should be moved by a mild cathartic, and alkaline drinks used.

A teaspoonful of common pearlash or soda in a tumbler of water, will suffice for a day. Locally, the best application is powdered chalk, scattered freely and frequently over the reddened surface.

The trouble is generally merely a symptom of disorder of the stomach and liver, and the treatment therefor must be directed by the physician to the improvement of the condition of the whole system. Just as soon as the inflammatory attack begins to subside, and the tongue to become of natural color, there is need of tonics, of which the best is substantial food, though even stimulants may sometimes be required.

It is more and more regarded as betokening a want of tone in the system, and as demanding a sustaining treatment.

Where the head is involved, or where the tissue beneath the skin is much affected, early medical advice will be required.

SCABIES OR ITCH.

This to all is a very affecting disease, and no one who has it but regards scratching as a poor and unprofitable way of making one's living. It is caused by an insect, whose favorite locality is near joints and crevices. This, different from most skin diseases, is entirely local in its action; and even when treated constitutionally, the only design is to produce such general effect as shall destroy the insect. The intense itching, and the formation of small vesicles or eruptions filled with watery fluid, especially between the fingers and about joints, but never on the face, are distinctive enough of the disease. The vesicles at first are pointed and surrounded by a red margin, and after by irritation become filled with matter. The disease is communicable only by the insect, and is therefore never epidemic, but caught by direct contact. It may appear in three days after exposure, or not until several weeks.

The treatment of the disease is generally simple, direct, and effective. Take a teaspoonful of flower of sulphur, half a teaspoonful of saleratus, and a tablespoonful of lard, mix them, and apply the ointment on the part affected every morning and night until the disease is removed. Common salt, in double the quantity, may be substituted for the saleratus. Where the affection is extensive or of long standing, the entire body should be anointed. A warm bath should be taken every other day, and the treatment will usually be completed in five days. The clothes worn before should be suspended in a hot oven, and then exposed to the air, so that the insect may be destroyed, if there be any clinging thereto. To most, however, the use of the sulphur ointment is unpleasant, as it advertises the disease. Where some other method is preferred, soaking the part in strong brine every six hours will often cause the death of the insect. A strong tea or infusion of thyme, to a pint of which a tablespoonful of alcohol, a few drops of oil of peppermint, lemon, turpentine, and vanilla have been added, will also have the same effect in a few days. The following forms a very good, effectual, and cheap liniment:

Soft soap,	.	.	.	one tablespoonful.
Alcohol,	.	.	.	one teaspoonful.
Vinegar,	.	.	.	" "
Salt,	.	.	.	" "
Chloride of soda,	.	.	" "

We find then that, notwithstanding general impression, this is not so unmanageable a disease, but that something besides fire and brimstone will cure it.

ECZEMA, IMPETIGO AND PRURIGO.

It is too often the case both with patients and some physicians, that independent of the eruptive diseases already mentioned, they have but three other classes, calling all smooth red spots without any discharge, erysipelas, and all others tetter, or salt rheum. Now the true character of erysipelas, we have already defined. As for the term tetter, it is about as descriptive as the term breaking out, and though it sounds a little more scientific,

Webster cannot trace its etymology, and medical writers vary in its application. Chronic eczema is one of the most frequent diseases known by this name, and the term tetter is perhaps more commonly applied to it than to any other.

Impetigo is another disease, differing from the former, to which this same name is also frequently given. Prurigo is still another affection, called also in common, tetter, and the seven years' itch. We purpose to notice these diseases together, in order that we may more definitely mark the distinctions between them.

I.—Eczema is a vesicular disease; that is, if you notice the pimple before it is broken by scratching, it is filled with a fluid.

Impetigo is a pustular disease from its very commencement, the points of eruption being filled with matter. In prurigo, the eruptive point is filled neither with water or matter, but is a solid elevation above the surface.

II.—Eczema generally passes from an acute to a chronic stage, being at first attended with heat and fever.

Impetigo is generally accompanied with less acute symptoms, but the discharge from it is more plenteous, and dries more thickly upon the skin.

Prurigo is a chronic disease, seldom accompanied by any acute symptoms.

III.—Eczema, if called tetter at all, must be called wet tetter; but as it becomes dry, a thin, scaly crust forms, like loose scarf-skin.

Impetigo, on the other hand, forms a thick crust of

scabs, and much of what is called scald-head, is this disease. It quite frequently makes its appearance upon the faces of children, and, when the scabs are removed, leaves the surface for a while red and discolored.

Prurigo has no discharge or crust whatever, except such as results from the elevation of the skin, or from the scratching to which the intolerable itching gives rise. This latter sometimes causes a black scab, made up of the little blood which has exuded.

IV.—Eczema, though not contagious, may be caught by direct contact of surfaces, as by sleeping in the same bed.

Impetigo is not communicable.

Prurigo is regarded as not contagious, but may, I think, be transmitted, by direct and frequent contact.

All of these diseases are accompanied with itching. Of these, that of prurigo is the most persistent and intense, and that of impetigo the least. An Irish friend of mine, being informed by the doctor that his child had the croup, replied, "An sure, doctor, I don't know what he's got, but whatever it is, he's got it fuss rate." In this sense, the itching of prurigo may be said to be "fuss rate," but either of the other two are abundantly sufficient.

In eczema, we regard the best external treatment to be, the plentiful and frequent application of prepared chalk. This is valuable as an alkali, and by absorbing the moisture, does much to control the uneasiness and check the progress of the disease.

Internally, a seidlitz powder, or half a tablespoonful of cream-of-tartar every other morning, and the intermediate morning a half thimbleful of soda or saleratus, dissolved in water, and given a half hour before eating, will often be of service. We believe injury to be often done by the use of wet appliances and baths, early in the disease, except so far as these are necessary to promote cleanliness.

If these means, persevered in, do not succeed, the tincture of Spanish flies or Pearson's solution probably will, but these should be given only by the direction of the medical attendant.

Impetigo, if treated at once, upon its appearance, requires, externally, the chalk powder, as eczema, but, as the discharge from it is greater, and more disposed to form in large thick crusts, frequent warm baths and poultices may be required. The design of these is to remove the matter, and this once done, poultices must be dispensed with, and drying-powder again employed. It is sometimes necessary, when the disease does not yield, to use some medicated lotion. Of these, one of the best is, a teaspoonful of soda and a scruple of acetate of zinc, in a pint of flax-seed tea. After the use and drying of the wash from the part, the chalk should be re-applied.

Prurigo, even when treated early, is likely to be tedious and troublesome. Internally, the use of alkaline drinks is indicated. A half-teaspoonful of saleratus each morning, in water, is among the best. Externally, we know of no better treatment than that mentioned for the itch, except that washes are preferred to ointments, and the materials recommended should be used mingled with water rather than lard.

Each of these three diseases will, in severe cases, need the advice and treatment of the skilled practitioner, and even he will often be compelled to resort to one remedy after another, in the hope of restoration.

For relief from the tormenting itching, which at times occurs in all these diseases, we know of no preparation better than that recommend by Wilson.

Corrosive Sublimate, 6 grains,
Spirits of Rosemary, } Of each 1 ounce,
Spirits of Wine,
Emulsion of Bitter Almonds, 6 ounces,
and applied frequently.

BARBER'S ITCH, NETTLE RASH, &c.

There are various other skin-diseases, a full description of which is not necessary in a book like this, but in respect to which some valuable hints may be given in general.

Mentagra, Chinwhelk, or *Barber's Itch*, is a pustular disease, located in and about the beard, and capable of being transmitted by contact. The razor must be laid aside for the scissors, and if there is much pain, redness and swelling, the flax-seed and slippery elm poultices must be used until inflammatory action is abated. A weak wash of saleratus will then be of service. In respect to the system at large, the indication is to keep the bowels in a laxative condition, and to avoid pork and greasy food of any kind. Where the eruption still continues, recourse must be had to such soothing or stimulating applications as the medical adviser may suggest, and if you get over it in six months be thankful.

Urticaria, or *Nettle Rash*, is almost always dependent upon disorder of the stomach, and is frequently the direct result of eating lobster, crabs, salted meat, or other indigestible food. It usually appears in blotches, with hard edges, and slightly raised in the centre, like a sting or a risen mark from the end of a whip. The itching is often most intense. The blotches will disappear, and then reappear, and we have often seen the associated with intermittent fever. If the attack ; caused by recent food, an emetic of ipecacuanha will afford the most speedy relief. Where it is the result of more permanent disorder of the digestive organs, other remedies will be required. Antacids, as soda or lime-water are generally indicated, and the snake-root tea has been regarded by some as specific. It is sometimes found, that some one article of food has been the unsuspected cause of the eruption, and this discontinued, all trouble has ceased. All internal remedies prescribed must be directed to the condition of the stomach and liver, while the best local applications are those already mentioned as tending to control local uneasiness.

Herpes is another skin affection, showing itself variously, in the form of ring-worm, fever-blisters, so called, on the lips, and shingles, which by some strange notion, have the reputation of being fatal if they encircle the body. The treatment, when any is needed, will differ but little from that of the last-mentioned ailment.

Psoriasis, commonly known as *Salt Rheum*, is among the most common forms of chronic skin-disease. It ap-

pears in patches on different parts of the body, has a
rough feel to the touch, and is covered with scaly
whitish skin, which, from time to time rubs off like particles of bran.   Often a single spot of it will last for
years, or it may diffuse itself more extensively. Arsenic
is the best remedy for this disease, not in very large
doses, for though this would terminate it, the cure
might not be acceptable, but taken in small quantities
under the direction of the physician, it is both harmless and unattended with any serious after results.   It
is the only reliable remedy for this affection, and it may
be used both externally and internally and properly administered, seldom fails to afford relief.

We have already referred to a sufficient number of
these skin affections to show that they are not all tetter,
and many more might be enumerated, but the rest are
so rare or need so universally the careful discrimination
of the physician, that it would be useless to describe
them here.

ASTHMA.

Asthma is a spasmodic disease of the lungs, the
prominent symptom of which is great difficulty of
breathing.   We meet with it both in a chronic and
acute form.   In the former the patient is often affected
even by slight exertion, and certain conditions of the
atmosphere especially predispose to an attack.   The
symptoms sometimes supervene very suddenly, the person being roused from sleep gasping for breath, and im-

mediate suffocation seeming inevitable. Sometimes, while walking leisurely along without any apparent cause, there will be a sudden oppression. The effect of the spasmodic action is much the same as that produced by smothering or by any thing preventing the free access of the air to the lungs. The disease often seems to be hereditary. In asthma the great distress is in the expiration or flowing out of the air, which distinguishes it from mere oppression. The remedies first to be used are those which have a tendency to allay spasmodic action. The following is a very good mixture, to be kept constantly on hand by any person subject to these attacks :

Compound spirits of lavender,
Tincture of valerian,
Paregoric,
Sulphuric ether,

of each equal parts.

In acute cases a teaspoonful may be taken every half hour until relief is obtained, and in milder cases one every three hours. A hot foot bath, and a mustard plaster over the chest, are always of benefit. Persons subject to this disease should be careful by the use of flannel, or by other means to keep up an equable temperature of the skin. Excitement should be avoided; exercise taken regularly, but not excessively, and the stomach should never be overloaded. When the severity of the spasmodic action has been overcome, expectorants will be of advantage, and of these, that named for hooping cough is one of the best. The frequent use of the bath, especially of the shower bath, has sometimes been found to fortify the system against the disease.

DYSPEPSIA.

Derangements of the stomach differing very much in their character, and in the treatment required for them are designated by this term. When appetite fails, or when the food which should nourish, and sustain, produces pain, wasting and discomfort, the sufferer at once feels that the main spring of life is ajar, and that means must be sought to restore vigorous action. It is the special, and peculiar province too of dyspepsia to bring with it a dire train of nervous sensations, and imaginary evils, and the worst cases of hypochondriacal foreboding, and sorrow, are not unfrequently dependent upon some failure on the part of the stomach, properly to dispose of the materials which enter it. All rules, and directions having reference to the recovery from dyspepsia, have their starting point in a proper regulation of the diet, and it is almost in vain to attempt medication unless the sufferer has that self control which will enable him to eschew instead of chew articles of food, which are known to be deleterious, and to lead him both as to quantity, quality, and various other points, to be obedient to the indications of his disease. It is seldom that the dyspeptic needs to abstain much from food, but there are certain articles which must be omitted, and certain rules which must be observed, or else all the physic of the most skilled would be better cast unto the dogs. In the article upon diet, points have already been noticed which have a direct bearing upon this ailment, both as to its prevention, and cure, and a careful notice has enabled me to find very few cases of dyspepsia among those who exercise regularly, eat moderately, chew thoroughly, and observe system in all those functions of life which are under the control of the will. That keen observer, William Hunter, exhibited his admirable good sense in his treatment of a case, an account of which we here transfer, mainly in his own words, from one of his volumes of Medical Observations and Inquiries:

"Many years ago a gentleman came to me from the eastern part of the city, with his son, about eight or nine years old, to ask my advice for him. The complaint was pain in the stomach, frequent and violent vomitings, great weakness, and wasting of flesh. I think I hardly ever saw a human creature more emaciated, or with a look more expressive of being near the end of all the miseries of life. The disorder was of some month's standing, and from the beginning to that time had been daily growing more desperate. He was at school when first taken ill, and concealed his disorder for some time; but growing much worse he was compelled to complain, and was brought home to be more carefully attended. From his sickly look, his total loss of appetite, besides what he said of the pain which he suffered, but especially from his vomiting up almost everything which he swallowed, it was evident that his disorder was very serious.

"Three of the most eminent physicians of that time attended him in succession; and tried a variety of med icines without the least good effect. They had all, as the father told me, after sufficient trial, given the patient up, having nothing further to propose. The last pre-

scription was a pill of solid opium; for in the fluid
state, though at first the opiate had stayed some time
upon his stomach, and brought a temporary relief, it
failed at length, and like food, drink, and every medicine which had been given, was presently brought up
again by vomiting. The opiate *pill* was therefore given
in hopes that it would elude the expulsive efforts of the
stomach. It did so for a time; but after a little use,
that likewise brought on vomiting. Then it was that his
physician was consulted for the last time, who said that
he had nothing further to propose.

"Though at first the boy professed that he could assign no cause for his complaint, being strictly interro-
gated by his father, if he had ever swallowed anything
that could hurt his stomach, or received any injury by
a blow, or otherwise, he confessed that the usher in the
school had grasped him by the waistcoat, at the pit of
the stomach, in a peevish fit, and shaken him rudely,
for not having come up to the usher's expectation in a
school-exercise, and that, though it was not very painful
at the time, the disorder came on soon after. This account disposed the father to suspect that the rude grasp
and shake had hurt the stomach. With that idea he
brought him to me, as an anatomist, that an accurate
examination might, if possible, discover the cause or
nature of the disorder.

" He was stripped before the fire, and examined with
attention in various situations and postures; but no fullness, hardness, or tumor, whatever, could be discovered;
on the contrary, he appeared everywhere like a skeleton
covered with a mere skin; and the abdomen was as

flat, or rather as much drawn inward, as if it had not contained half the usual quantity of bowels.

"In the adjacent room, I said to the father, 'This case, sir, appears to me so desperate, that I could not tell you my thoughts before your son. I think it most probable, no doubt, that he will sink under it; I believe that no human sagacity or experience could pretend to ascertain the cause of his complaint; and, without sup posing a particular or specific cause, there is hardly anything to be *aimed at* in the way of a cure. Yet, dreadful as this language must be to your ear, I think you are not to be without hope. As we do not know the cause, it may happen to be of a temporary nature, and may, of itself, take a favorable turn ; we see such wonderful changes every day, in cases that appear the most desperate, and especially in young people. ·In them the resources of nature are astonishing.'

"Then he asked me if I could communicate any rules or directions, for giving him a better chance of getting that cure from nature, which he saw he must despair of from art.

"I told him that there were two things which I would recommend. The first was not so important, indeed, yet I thought it might be useful, and, certainly, could do no harm. It was, to have his son well rubbed, for half an hour together, with warm oil and a warm hand, before a fire, over and all around his stomach, every morning and evening. The oil, perhaps, would do little more than make friction harmless, as well as easy ; and the friction would both soothe pain and be a healthful exercise to a weak body.

"The second thing that I had to propose, I imagined to be of the utmost consequence. It was something which I had particularly attended to in the disorders of the stomach, especially vomitings. It was, carefully to avoid offending a very weak stomach, either with the quantity or quality of what is taken down, and yet to get enough retained for supporting life. I need not tell you sir, said I, that your son cannot live long without taking *some* nourishment; he must be supported, to allow of any chance in his favor. You think that for some time he has kept nothing of what he has swallowed; but a small part must have remained, else he could not have lived till now. Do you not think, then, that it would have been better for him, if he had only taken the very small quantity which remained with him and was converted into nourishment? It would have answered the end of supporting life as well, and perhaps have saved him such constant distress of being sick and of vomiting. The nourishment which he takes should not only be in very small quantity at a time, but in quantity the most inoffensive to a weak stomach that can be found. Milk is that kind of nourishment. It is what Providence has contrived for supporting animals in the most tender stage of life. Take your son home, and as soon as he has rested a little, give him *one* spoonful of milk. If he keeps it some time without sickness or vomiting, repeat the meal, and so on. If he vomits it, after a little rest try him with a smaller quantity, viz., with a dessert, or even a teaspoonful. If he can but bear the smallest quantity, you will be sure of being able to give him nourishment. Let it be the sole business of one person to feed him. If you succeed in the beginning, persevere with great caution, and proceed very gradually to a greater quantity, and to *other* fluid food, especially to what his own fancy may invite him, such as smooth gruel or panada, milk boiled with a little flour of wheat or rice, thin chocolate and milk, any broths without fat, or with a little jelly, or rice, or barley in it, &c., &c.

"We then went in to our patient again; and that he might be encouraged with hope, and act his part with resolution, I repeated the directions with an air of being confident of success. The plan was simple and perfectly understood. They left me.

"I heard nothing of the case till, I believe, between two and three months after. His father came to me with a most joyful countenance, and with kind expressions of gratitude told me that the plan had been pursued with scrupulous exactness, and with astonishing success; that his son had never vomited since I had seen him; that he was daily gaining flesh, and strength, and color, and spirits, and now grown very importunate to have more substantial food. I recommended a change to be made by degrees. He recovered completely, and many years ago he was a healthy and a very strong young man."

The bundle of excellent good sense which this quotation from Hunter, contains, so applicable in principle to manifold other cases, will excuse its length.

A similar pursuance of the dictates of common reason, would cure many a case of confirmed dyspepsia; but, alas! no severer task does the physician undertake, than that of attempting to control the appetites of his patients. Disorders of the stomach indicate themselves in varied and changeable ways. In some, there is a sense of pain after eating; in others a mere fullness, or sense of oppression; in some, belching of wind and other signs of sour stomach; in others, nausea and vomiting of food. Now and then there is almost an entire want of appetite, and even an actual loathing of food; but more frequently the hunger is variable—at one time voracious, at another easily satisfied.

Heart-burn, water-brash or the raising of the food unchanged, are in other instances the chief symptoms. The causes of these different symptoms may be either mere debility of stomach, or an absence of sufficient gastric juice, or ulceration, or cancer of the stomach; in fact there are manifold forms of disease, some mild, some most severe, of which these external and sensible signs are but the telegraphic dispatch.

Want of appetite is usually dependent upon want of tone in the digestive organs, and is generally relieved by the use of mild stimulants or tonics. It is in this class of cases that the extracts of the bitter herbs are valuable, and wormwood, columba, gentian, or quinine, often cause speedy relief. If constipation exists, this must first be relieved by a mild cathartic. Where heart-burn and water-brash are prominent symptoms, acidity is the usual cause, and soda, saleratus or lime-water will relieve for the time, while any of the bitters above named will help to overcome the cause of this unnatural acid secretion.

If there is vomiting after meals, prussic acid prescribed in mild doses by the physician, will, after a time, often afford relief. In some cases there is a demand for acids, instead of alkalies, and of these lemon-juice is one of the best. Dyspepsia, in most of its forms, is apt to be tedious and subject to relapses, and hence its treatment will often need such variation as a careful medical observer will indicate. But the great point to impress upon the patient, is the necessity of obeying the laws of temperance and of health, and to regard medicine as merely auxiliary to these other and more important means.

THE HAIR.

This, nature's beautiful and appropriate covering for the head, is an ornament which all are disposed to retain as long as possible, and yet one which not unfrequently is injured or removed by disease or neglect. The hair-bulbs are situated in a looping down of the skin, and the hair is lengthened by a constant growth therefrom. In order that it may be kept in a healthy state, considerable attention thereto is necessary. By want of care the oily secretion by which it is kept glossy is obstructed; by dandruff or other diseases of the head, the roots are injured, and sometimes the bulb itself destroyed, so that any thing which has a tendency to cause an unnatural state of the skin of the scalp will, in time, affect the hair. Interference with natural perspiration and exhalation will often do this. Warm caps and close hats are injurious both to the head and hair. The perspiration constantly being given off from the surface, is thus confined, and cannot but give rise to an unhealthy condition of the part. Caps or hats should never be worn without a ventilator in the top. These actually keep the head more comfortable, and prevent those colds which are often contracted by the sudden removal of the close hat, and the consequent exposure of the head to a different temperature.

The frequent use of the fine comb often causes irrita· tion of the scalp, which results in the formation of dandruff. Its only use, therefore, is gently to remove all foreign substances from the head, and the severe combing which often injures both the hair and the temper of the child, should be avoided. The use of a harsh brush, as in shampooing, is far preferable. The hair of children, both of boys and girls, should be kept moderately short; not so short, on the one hand, as to cause a bristle like stiffness, nor so long as to cause it to become dry and discolored. If too thick, it interferes with the health of the surface of the head, and often causes too great heat, and consequent headache and uneasiness. Frequent wetting of the hair with cold water, though sometimes beneficial to the head, is not a good habit, so far as the healthy condition of the hair is concerned; and where it is necessary, special care should be taken to prevent continued moisture. Where there is an absence of natural oily secretion, the use of any pure oil, in very small quantities, will often be found beneficial. Good fresh butter, or sweet oil, scented with the oil of bergamot, in the proportion of a drop to a tablespoonful, is much better than bear's grease or manufactured pomatums. If there is dandruff on the head, and the hair is dry and unhealthy, you cannot use a better preparation for it than the following:

95 per cent. alcohol, . . . 1 pint.

Tasteless castor oil, . . . 2 gills.

Tincture of Spanish flies, prepared
with 95 per cent. (pure) alcohol, . 1 gill.

Oils bergamot and wintergreen, . 15 drops each.

This well shaken and applied to the hair from time to time, will be found much better than a pure oil, imparting to it a silken glossiness, and removing and preventing the constant accumulation of the dry scales of dandruff.

THE TEETH AND TOOTHACHE.

The preservation of the teeth is important, not only on account of their beauty, but also of the important office they perform in connection with the process of digestion. The careful mastication of the food is an indispensable part of the preparation which it needs before it is conveyed to the stomach, and by no other means can this be accomplished as well as by natural sound teeth. These commence in the form of small cellules on the mucous membrane of the jaw, and according to specific laws, become afterward developed into teeth. The tooth presses its way in the process of its growth until it cuts through the membrane above, and appears on the outside surface of the jaw. This in children often causes irritation of the nervous system

10*

and of the mucous membrane. If the process is retarded and the gum much swollen, lancing the gum down to the tooth is advantageous, as it removes the pressure, and by permitting the escape of a little blood, relieves the local inflammation. It is a mistaken idea that if this heals again before the tooth is through, that the process is more difficult, for even the cicatrix or healing point is less vascular, and more easily penetrated than it was before.

Children at an early age should be taught to pay good attention to the teeth. The first set is often neglected, on the ground that it is to be replaced, but this is mistaken policy. Decaying teeth injure the stomach, and locally the gums, so that the surface of the jaw is illy prepared for the reception of new and perfect teeth, and it is even believed by some, that the decay of the first teeth is transmitted to the second. Be this as it may, a decayed set of first teeth are almost certain to be succeeded by those which become poor early in life. The great cause of decaying teeth, is the presence of small particles of food between them. These are retained in the small crevices, or between adjacent teeth, until they undergo such processes as generate acid and substances injurious to them, and thus points of decay are commenced, or a roughness of the dentine occasioned, to which deleterious materials adhere. The habit of rinsing the mouth after each meal, is an excellent one, and is easily taught to children of medium size. If the teeth are kept clear of any animal or vegetable substance between them, and the stomach kept in such a state that the breath is pleasant, you need not fear their decay.

It was never intended that the masticators should be gone while there is need for their use; and false sets are not only troublesome and expensive, but are too often the signs of early neglect. We scarcely realize how much the state of the teeth has to do with that of the stomach. An unhealthy condition of the mouth, has frequently been the moving cause of dyspepsia, and disorder of the stomach, in its turn, often declares itself by premature decay of the teeth. Attention on the part of parents will do very much to prevent either of these. If, even with frequent rinsing, there is any collection upon the teeth of the child, a little powdered myrrh or charcoal may be rubbed upon them, by means of a wet rag over the finger, and then cold water used for rinsing as before. If there is still a tendency to an accumulation of coating, a brush should be used. We are persuaded, that did parents generally realize the subsequent inconvenience, expense, and sometimes ill-health, resulting from the neglect of these simple means, they would direct more attention to the preserving of the teeth of their children. In these days of sugar piece-meals, confectionery and carelessness, we too often find the voice of the grinders low in youth, instead of ceasing only with declining age.

The first teeth usually make their appearance between the sixth and seventh month, but they often vary much from this, and longer delay need excite no anxiety. The first permanent teeth are cut about the sixth or seventh year. The number of first or milk teeth is twenty, of the permanent thirty-two.

The following is a statement of the order of time in

which both the first and second sets usually appear; but there is so much variation in special cases, that it is impossible to give an accurate law. The lower teeth of each kind generally precede the upper.

TEMPORARY TEETH.

The 4 middle incisors, or first teeth, two
  below and two above, ... 7th month.
4 side incisors, or front teeth, one each side
  of the first, below and above, . . 9th month.
4 first molars, or back teeth, . 12th to 15th month.
4 canine teeth, i. e., the stomach and eye
  teeth, between the side incisors and
  molars, below and above, one each side, 18th month.
4 second molars, or back teeth, below and
  above, ... 2 years.

Before these teeth are shed, the jaws have elongated with the growth of the child, and hence a greater number will be required to fill up the vacancy. This is supplied by what are called the first and second bicuspids, which, unlike the sharp-pointed front teeth and four-cornered molars, have two sharp surfaces, with a line of depression between them, and hence called bicuspids.

When the growth of the jaw is completed the last vacancy is supplied by the third molar or wisdom teeth.

PERMANENT TEETH; OR, RATHER, THE SECOND SET.

First molars of the second set of teeth,	6½ to	7th year.
Central incisors, or middle front teeth, . .	7th	"
Lateral incisors, ...	8th	"
Eye and stomach teeth, ...	9th	"
Second bicuspid, ...	10th	"
Second molars, or back teeth, .	12½ to	13th "
Third molar, or wisdom, . . .		17 to 25th "

If the first teeth are much decayed, they should be removed to make way for the others, not because the others are not able to make way for themselves, but to prevent their being affected by their predecessors. Sometimes, it is true, it may be necessary to extract one of the first teeth, to insure evenness in the second set, but nature has abundantly provided the means of getting rid of the first, except in rare instances.

If a tooth has been cut out of position, much may be done to remedy the evil, by daily pressure upon it, as it is not at first so firm in its socket but that it may be pushed into the proper line of symmetry. The permanent set having made its appearance, the great desideratum is, to know how to preserve it; for comfort and economy unite in prompting us to take good care of this portion of our frame-work. If not, through long torments of toothache, neuralgia, and the like, we will too soon be rid of them, to the damage of health, pocket, personal appearance, and at the expense of much pain.

Consider it, then, as a part of your daily business to attend to the teeth. The mouth should be cleansed after each meal, and any meat fibres remaining between the teeth should be removed. As many substances eaten have a tendency to adhere to the teeth, and are not removed by rinsing of the mouth, a brush should be used each day, to prevent any accumulation. The harshness of the brush and the amount of rubbing required, are very different for different persons, and are to be regulated by the result. If a soft brush, with clean cold water upon it, will prevent any tartar forming upon the teeth, this alone is best. If not, a more firm bristle must be used, and a tooth-wash or powder, which, by the friction of its minute particles, may remove the offending material.

If any thing is permitted to accumulate upon the teeth, the enamel, or hard outside covering of the tooth, is penetrated, and this once gone, the decay of the remainder is rapid. If any tartar has already been deposited, you should choose the least of two evils, and have the dentist remove it, and then prevent its regathering by use of the brush.

As soon as the slightest cavity in a tooth is discovered, it should be filled, before the inner portion becomes affected. Thus toothache will be prevented, and the tooth preserved. Sometimes, as we have hinted, independent of the evil arising from fermenting, decomposing particles of food between and about the teeth, they are injured by a bad condition of the stomach. A foul breath and a coated tongue always produce some impression upon the teeth. Here something more than rinsing and friction is indicated. The object should be to overcome the state of the system by corrective medicines, and to sweeten the breath by antiseptic gargles, or washes, used in the mouth and upon the teeth. Among these, charcoal, myrrh, chloride of soda and prepared chalk, rank as the most efficacious. The following tooth-wash we regard as the best preparation for the teeth that can be devised :

Prepared chalk . . . . 1 ounce.
Chloride of soda . . . 2 drams.
Tincture of myrrh . . . 1 ounce.
Pure water . . . . 4 ounces.
Oil of wintergreen . . . 3 drops.

This should be shaken before using, and a half teaspoonful each day is sufficient. The teeth are often materially injured by the use of very hot or cold drinks, the enamel thereby being cracked, and in time the inner portion of the tooth exposed.

Toothache is dependent upon various causes, such as derangement of the stomach, sudden cold, neuralgia, an unhealthy state of the gums, corroding of the outer portion of the tooth, and exposure of the nerve. Where, as is not unfrequently the case, the cause is stomach disorder, a mild cathartic will often give relief, and a Seidlitz powder before breakfast, and at night before retiring, will very frequently cure the toothache. If you have taken a severe cold, apply the remedies for cold, as a warm general or foot bath, a good bowl of some hot herb-tea, and thus restore the action of the skin. If there is neuralgic affection, an opiate, as thirty drops of laudanum for an adult, will often procure relief. Where there is decay of the outer portion of the tooth, local appliances are of more service. Holding in the mouth some strong hop-tea, ap-

plying a mustard plaster, and after that, a hop poultice
to the face, and rubbing the region about the tooth with
any essential oil, as cloves, origanum, or cajeput, or with
a little chloroform, will often lull and abate the pain.
A general or foot bath should not be omitted. When
there is an opening, made by decay, leading to the
nerve, the indication is to address our remedy to that
particular point. The opening should be carefully
cleansed and dried, which may be done by a bit of cotton on the end of a long needle or piece of wire, and
then a small piece of cotton moistened with laudanum
and sweet oil, or with any of the appliances above
named, should be introduced into the aperture. Only
enough should be used to close the opening and convey a little of the sedative to the part, for if the hole is
pressed very full, the nerve will be still more affected.
If these means do not succeed, we know of no better
advice than that you should hasten to present yourself
before some one skilled with the forceps, and an adept
in the art of extraction.

POISONS.

These are sometimes taken by mistake, or by those
who repent of their attempt at self-destruction as soon
as it has been made, and life frequently depends upon
the immediate use of means which, if delayed, will
prove of no avail. Chemistry has provided for many
of the most active poisons equally active antidotes;

and, as you sweeten sour food by soda, so you may often change the character of a poison taken into the stomach, by knowing what substance will make with it a new and harmless compound.

There are a few indications which apply in general to all cases of poisoning, which may be noticed before we proceed to specify each particular kind.

I.—Let the poison already in the stomach be diluted by liquids. Many poisons do their harm by their immediate action upon the coats of the stomach, burning the surface, or even eating an opening through its walls. Now, if abundance of liquids or soft food is quickly swallowed, this may prevent much of the injury. Many of these articles, though fatal when concentrated, do no harm when diluted. Cases are on record of persons who as a curiosity have at once swallowed quantities of poison sufficient to kill many, but who by having the stomach beforehand filled with oil or watery fluids have suffered no inconvenience. Acids act so rapidly that too often any thing taken afterward is not quick enough to prevent the evil, but by checking it, even in this class of cases, food may be of avail. Warm water or milk are usually at hand, and may be administered in large quantities.

II.—If possible, the poison should be removed from the stomach. To this end vomiting should be caused as speedily as possible, and those emetics should be used which will most quickly excite this. A teaspoonful of mustard in a tea-cup of warm water, is generally the most ready, and may be given to an adult, or half the quantity to a child every ten minutes, till its effect is produced. A teaspoonful of powdered ipecac, if at hand, given in the same way, will act as well. After either has been given, passing the finger down the throat, or tickling about the tonsils with a feather may hasten the effect. Any emetic at hand may be employed in small doses at repeated intervals, but the two mentioned are the safest for the friends to prescribe.

III.—The stomach should be protected by mucilaginous or oily drinks. Flax-seed or slippery-elm tea, or eggs or jelly, or a tablespoonful of oil, or melted butter or lard, or molasses may with propriety be given. These while not interfering with other remedies serve to protect the surface of the stomach.

IV.—The chemical antidote should forthwith be used. This is better than any thing else, because it absolutely kills the poison, and while other means should not be omitted, this is the most certain remedy. Thus, if an acid has been taken, soda, saleratus, lime, magnesia, or prepared chalk should be mixed with water, and given in frequent doses. . Of these, the best is the calcined magnesia given freely. If an alkali has been swallowed, as a lump of potash or lime, then acids, as vinegar, cider, lemon juice, and the like are indicated; but the use of oily, and mucilaginous drinks must not be omitted.

V.—Whenever from poisoning there is stupor, or a tendency thereto, cold to the head, sprinkling cold water frequently in the face, warmth to the extremities, and hot coffee internally, are always appropriate.

VI.—As soon as the accident is discovered, the physician should be sent for, that he may be on his way while these means are being used, for after all vomiting may not be secured, and the stomach-pump may be needed, or some antidote with which he is acquainted; and even where first symptoms are overcome, treatment is often required to prevent future trouble. With these general points to guide the friends of the sufferer, we next notice the more usual poisons, and the special means to be applied in overcoming their effects. One of the most common is

I.—*Corrosive Sublimate.*—This is an intensely irritant poison, two or three grains of which often cause death. It has much the appearance of white loaf-sugar, pulverized, but has an intense, coppery, metallic taste, which you may detect without danger by taking up a little on the head of a wet pin and tasting it. The best antidote for it is uncooked fresh eggs, of which a few should be immediately beaten up and administered. Warm drinks should also be given, and effort be made to induce vomiting. If severe symptoms continue, eggs may be given every fifteen minutes; but vomiting should be secured, as the new compound formed will sometimes be redissolved, if not removed. If eggs are not at hand, rye or wheat flour, made into a thin paste with milk, may be freely given. Lime-water or soap-suds, in small quantities, will also be advantageous.

II.—*Arsenic.*—With this, one of the most common and virulent of poisons, we are familiar under two forms —one the pulverized steel gray fly-powder, of the shops, the other the white arsenious acid usually known simply as arsenic, and used both for the destruction of lower and higher animals. Less than four grains have caused death, while on a full stomach, with vomiting, soon after, ten times the quantity has produced little effect. The powder may be known by its weight, its pure, soft • whiteness, like that of the best white indian meal, and by its sweetish, astringent taste, which you may test by tasting. If you will dissolve some of it in rain water, then add a little lunar caustic, afterward some soda, yellow globules will form, and you thus have a chemical test. Burning heat at the region of the stomach, violent retching, great thirst and dryness about the throat, and symptoms of inflammation of stomach and bowels, accompanied by great prostration, are the usual effects of this poison. No true antidote for it is generally at hand. Fresh calcined magnesia, in milk, may be given by the tablespoonful, and this sometimes seems to take up the arsenious acid. Warm water, or milk, eggs, lime-water, or oils, may be freely used. With these the mustard emetic should be combined, and vomiting promoted. The best antidote is the hydrated peroxide of iron, freshly prepared, and the stomach-pump is the most certain method of emptying the stomach; and the physician, when he shall arrive, will judge of the necessity of employing these. Much care is necessary, lest even after the first symptoms are relieved, violent inflammation supervene.

III.—*Oxalic Acid*, or acid from sorrel, is used in small quantities to make an acid drink, and by the housewife in cleaning metals, but is in large doses a most severe poison, acting upon the sides of the stomach like other concentrated acids, and causing extreme debility, and death. In fatal cases, it has usually been taken, by mistake, for

POISONS. 237

Epsom salts, which it much resembles. It may be distinguished from it and other articles, by its intensely sour taste, without bitterness, by becoming of a reddish brown color; if a little black ink is poured upon it, or, if a little lime-water is added to a solution of it, a white powder falls.

The antidote is chalk, which should be freely given in water by the teaspoonful, according to the severity of the symptoms. Common or prepared chalk, or whiting, which is ground chalk, will answer every purpose. In the absence of these, any other *alkali* may be used. There is no need of emetics, and prostration must be guarded against.

IV.—*Prussic or Hydrocyanic Acid.*—This, in its concentrated form, is the most active and powerful poison known. One drop of the pure will cause death, and even the smell of it may cause convulsions and fatal results. Nevertheless, in a diluted form, it is valuable as a medicine—it exists in the laurel, and in the kernel of the peach, bitter almond, and other fruits, and imparts to them much of their peculiar smell and taste. The pure acid is never employed except in the room of the chemist, and the only cases in which we usually have ill effects from it are where an overdose of the medicinal acid has been taken, or where children have eaten so freely of bitter almonds, peach kernels, or laurel, as to be affected thereby.

Where it does not cause almost immediate death, spasmodic breathing, dilated pupils, an almost imperceptible pulse, and convulsions, are the symptoms we find.

These, together with the smell like that of fruit kernels, will generally enable us to detect the kind of poison. Although known as an acid, it is feebly so. Hartshorn or ammonia should quickly be applied to the nostrils. Cold water dashed at intervals over the face and chest, chloride of soda, or chloride of lime, moistened with vinegar, freely sprinkled about the room and near the patient; and if these do not avail, artificial respiration should be resorted to, as directed for the drowned. As soon as the patient can swallow, ammonia, hot brandy, or spirits of turpentine and camphor, mixed, may be administered. The best chemical antidote is that of the Messrs. Smith. Give a solution of twenty grains of carbonate of potash, which is about the same as a half teaspoonful of common pearlash, in water, and immediately after twenty grains of copperas, or a piece the size of a red cherry-pit, in a tablespoonful of water.

V.—*Sugar of Lead.*—Acetate or sugar of lead, from its sweet taste and resemblance to white sugar, is sometimes mistaken for it and swallowed by children, in their fondness for sweet things. *Carbonate of lead*, or *white paint*, and *litharge* and *red oxide of lead*, will, like the former, act as a poison upon the system. The symptoms are, great pain in the stomach, vomiting, and the formative symptoms of inflammation of the bowels. Effort should at once be made by mild emetics or cathartics, to remove the substance, and then to control its effects by the use of opiates. Epsom salts is both an antidote and cathartic, and therefore a good dose should immediately be given. Oil of vitriol enough to make a tumbler of water taste a little sour, may at once be prescribed.

If there are symptoms of colic, as griping pain and
constipation, castor oil, with twenty-five drops of laudanum in it, for an adult, may be administered from time
to time. Injections of flax-seed tea with laudanum in
each, and the warm bath, are often efficacious. Lead
poisoning sometimes occurs by small quantities gradually getting into the system and accumulating there, so
as to cause either local colic or paralysis. Workers in
lead, as painters or manufacturers, are particularly exposed, and water which has passed through lead pipes,
will in time affect many. Some are far more susceptible to this influence than others, and are very speedily
poisoned thereby. Some remarkable cases of its effect
in our own country have been registered in the report
of the American Medical Association, illustrating the
variety, and in some instances the severity of its effects.
In a family of five or six which became affected thereby,
under my own observation, one had dizziness of the
head and slight spasmodic action ; another a mere sense
of weariness and fatigue; another, a boy of ten, after
walking a mile from school, would sit down and cry
with the aching of his limbs, while one or two experienced no uneasy sensations. When it produces palsy,
as is sometimes the case, recovery is often slow and imperfect. All pure water acts on lead, and as some of
the compounds are dissolved or soluble, they are sus
pended in the solution. Even where the water which
has been standing in the pipe is all pumped out, still
the lower part of the tube is constantly in the water,
and its sides have been slightly acted upon, so that, we
believe, water should never be drawn through lead

pipes. The fact that all or most are not affected, is not
a conclusive argument, for the same would apply to
chills, the yellow fever, and the plague. Those who,
from their occupation, are compelled to be exposed to
lead in any form, or who are sickened by the smell of
paint, should drink five or ten drops of *diluted* sulphuric
acid in water, each day. The body should be frequently
and carefully cleansed, and if there is paint upon the
clothing, these should be removed after work is over.
The treatment of the different forms of lead-palsy is
such as to require the attention of the physician, and I
shall not, therefore enlarge upon these.

VI.—*Blue Vitriol* or *Blue Stone*, is a sulphate of
copper. Green Vitriol or *Copperas* is a sulphate of
iron, and *White Vitriol* is a *sulphate of zinc.*

Pure verdigris is an *acetate of copper. Litharge* is a
protoxide of lead, and red lead is a mixture of protoxide
and peroxide.

Saltpetre is a nitrate of potash. Any of these may be
taken in quantities which may prove poisonous, and
should be regarded as articles in respect to which care
should be used. The prominent symptoms from them
are · omiting, pain, and general irritation of the stomach
and jowels.

For poisoning with blue vitriol or verdigris, white of
eggs, or paste of wheat flour, should be freely given.
Flax-seed tea, made very sweet with sugar, is also of
service. For poisoning with the compounds of lead, the
best remedies are those indicated in case of sugar-of-lead
poisoning. When green or white vitriol, or yellow
ochre, have been taken in too large quantities, chalk

and flax-seed tea should be freely given. Any symptoms of inflammation must be promptly met by the medical attendant, and epsom salts or alum-water should be administered. For *lunar caustic*, which is nitrate of silver, common salt, given by the teaspoonful, is the antidote.

VII.—*Opium.*—This we are familiar with, in its solid form, as the dried juice of the poppy, and in its preparations, as morphine, laudanum, paregoric, elixir of opium, black-drop, and the like.

It is among the poisons most frequently taken, either purposely or by mistake, and the means to be used should therefore be well understood. We first remark, as a caution, that opium, morphine, or laudanum should never be given to a child under a month old by any one save the physician. Two or three cases have come under my own notice where death has been caused by a very small portion given by the nurse; a single drop even of laudanum has proved fatal, and hence it should never be used unless the medical attendant feels the exigencies of the case demand it. Three or four drops of paregoric will often relieve pain, and this is far safer.

The prominent symptom of an overdose of opium in any form is a tendency to profound stupor. The person falls into a deep sleep, the breathing usually slower than natural, sometimes accompanied with a snore; the skin is generally moist, the pupil of the eye contracted, the pulse small, and too often, in a few hours, the sleep ends in death.

Strong coffee is a valuable antidote in case of poison-

11

ing with opium. The great point is to keep the person
awake. Sprinkle water on the face, shake him about,
talk till he is tired of hearing you, walk him about the
room, and by every means in your power prevent sleep.
An emetic of mustard, and free application of it in the
form of plaster, externally, will be most proper.

Lime, tannin, or alum-water, may, with propriety, be
used, in conjunction with other means.

The spine should be rubbed with warm turpentine,
the extremities kept warm by bottles of hot water,
strong vinegar or hartshorn applied to the nostrils, and
thus every effort be made to keep up the sensibility of
the system, until the effect of the overdose has subsided.

Measure the activity with which you should apply
these means by the degree of stupor. If they do not
succeed, the physician may effect something by the
stomach-pump and stimulating emetics, or artificial
respiration may be resorted to, as directed in cases of
drowning. An occasional hot-bath is often of advantage, but it must not be continued long.

VIII.—*Nightshade, henbane,* and *thorn-apple,* or, as
the doctors call them, belladonna, hyosciamus, and stramonium, are three narcotics which, as common garden-
flowers, by which children are sometimes poisoned, and
as medicines which are frequently used, call for a passing
notice. They all belong to one botanical family and
class.

The belladonna may be known by its drooping purple
flowers, on short stems, and its small black and sweet
berries.

The hyosciamus or henbane may be known by its dark-yellow flower, streaked with purple veins, set close to the stalk, and its pod filled with many seeds.

The stramonium grows in rich grounds, as, near cattle-pens, and its rough, thorny apple, filled with kidney-shaped seeds, its large, erect, white, urn-shaped flower, and unpleasant odor, sufficiently designate it.

All these, when taken in poisonous doses, cause, at first, delirium and dilated pupils, but are quickly followed by profound stupor. For none of these is a specific antidote known, but the use of stimulating emetics, and of the other general means directed in cases of opium poisoning, are sometimes of service. Instead of coffee, mild acids, as lemonade, may be administered, and stimulants freely given.

Aconite, or monk's-hood, so called from the shape of its flower; the common hemlock, with its pretty white blossom, and the digitalis, or fox-glove, with its purple drooping flower, are also common garden ornaments, and used as medicines.

Not so powerful in narcotic effects as those above mentioned, they are, nevertheless, sedative in their action, and an overdose produces serious results.

The treatment for them is the same as that indicated for the last-mentioned poisons.

IX.—*Tobacco.* This is a sedative so frequently used as a luxury, that few have a correct idea of its intensely acrid and poisonous properties. When taken internally great prostration and depression are the results, and in large quantities convulsions and death follow its administration. Many are familiar with the effect produced when it is smoked or chewed for the first time, and although the system, by its habitual use may cease to respond to its power, it nevertheless often makes serious inroads upon the health, and causes disorder of the nervous and digestive functions. It contains an essential oil, a single drop of which will kill almost any of the smaller animals, and it is this which is deposited upon the sides of old pipes, and imparts to them their peculiar odor. The opinion of Dr. Warren, that cases of cancer of the lip occur most frequently in inveterate smokers, has been confirmed by other observers, and cancer and ulceration of the stomach sometimes seem to be the direct result of the use of the pipe. Smoking before eating is especially objectionable, since the weed acts more directly upon the coating of the stomach. It has been argued by some, that chewing is valuable because it preserves the teeth. The only way in which it can possibly do this is, by preventing the decomposition of particles of food retained between them, but as these are easily removed by rinsing the mouth, there is no need of resorting to such a roundabout preventive. On the hard surface of the tooth itself the tobacco exerts no influence, but the gums and surrounding portion, the health of which involve that of the teeth, are much injured by its action upon them. Not a single defence for the use of tobacco can be drawn from reason, experience, health or profit, and as a habit it does not even excite the admiration of its most devoted lovers. To some its use is, it is true, far more injurious and pernicious than to others; but the arguments for its employment by any one even lack the plausibility of those which may be adduced in favor of other stimulants or sedatives. In cases of slow poisoning by it, the great remedy is, to discontinue its use, and to have recourse to those remedies which invigorate the nervous system, and·give tone and energy to the digestive processes. Where, by any mishap, poisoning has resulted from its accidental or temporary employment, stimulation should be resorted to, both internally and externally, and a strong tea of white-oak bark or of tannin may be prescribed as a drink.

X.—*Camphor*, when taken in large quantities, produces serious symptoms; and internally it is not safe to be prescribed without special directions. I have known it in one case, where a patient took an overdose for pain in the stomach, to cause violent convulsions, and even without these there may be difficult breathing, cold sweats, paleness, and exhaustion. Cold should be applied to the head, a mustard plaster to the stomach, wine, brandy or ammonia should be given internally, and twenty drops of the tincture of Spanish flies with it, if at hand.

XI.—*Creosote*, easily known by its intense smoky smell, acts directly as an acid poison in the stomach, causing intense irritation, and affecting the nervous system. Even its use for decayed teeth is objectionable. White of eggs, if immediately taken, furnish ·the antidote to its effects, and a mild ipecac emetic, and mucilaginous drinks, as flax-seed tea, should be freely given.

XII.—*Chloroform* is an agent which has become so well known, that we have found it sometimes to be carelessly used by unprofessional persons. A case occurred to me not long since, in which, where I had prescribed it for a wash, the patient saw fit to smell of it for a nervous headache, and with but three or four inhalations became insensible. A few minutes after she awoke, perfectly relieved, but frightened when she remembered the cause. Too great care cannot be used with such agencies, and when, in the absence of a physician, any such effect is by mistake induced, hartshorn should be applied to the nose, water sprinkled in the face, and the extremities briskly rubbed. Artificial respiration is the chief reliance in stupor from chloroform.

XIII.—*Phosphorus*, even in small quantities, produces serious symptoms, and children have often been affected by swallowing the substance from the ends of matches. Burning of the mouth and stomach, excitement and convulsions, are the symptoms; and chalk or magnesia, given in plenty of any liquid, as milk or water, is the best antidote. Special symptoms must be met by special remedies.

XIV.—*Tartar emetic*, though a valuable medicine, may be taken in such quantities as to cause alarming symptoms. Its white powder, and its solution appearing like simple water, and even the taste not at first perceptible, it may be easily swallowed by mistake. The consequence is a metallic taste in the mouth, excessive and repeated vomiting, purging, cramping pains, and great prostration. Mucilaginous drinks should be given, and small quantities of laudanum and sweet oil, or melted lard, or a tea of white-oak bark or tannin, will be of great service. Injections of oil and laudanum are of benefit when the medicine will not be retained by the stomach ; and if the prostration is great, stimulation is also required.

XV.—There are a few common plants of a poisonous character, which are sometimes mistaken for harmless herbs, or eaten by children when unaware of their character.

The laurel, growing wild in our woods with its bright green leaves, and its numerous beautiful white flowers, if eaten in much quantity produces the symptoms resulting from an overdose of prussic acid, and the treatment indicated is precisely the same as therefor.

Those who cannot find enough of the good things of this life amongst the other provisions of nature, sometimes go out in search of *mushrooms* for breakfast, and mistaking the poisonous variety for those comparatively harmless, not a few have suffered therefrom. "The genuine mushroom is found in autumn, in rich old pastures. It has a small, round, brownish white head, of a delicate pink color underneath ; the stem is generally from two to three inches high. Such as are overtopped by trees or growing in the shade, must be avoided."

When the cut surface of a mushroom will tarnish a piece of silver held against it, you may be sure it is not fit to be eaten, and this is regarded by many as a test of its good or bad quality. The usual symptoms are those of colic, often followed by general cramp, disorder of the bowels, and even convulsions. The indication is at once by the use of mustard or other stimulating emetic to unload the stomach of its contents, and then prevent further trouble by the mild use of opiates, or of stimulants, if there be much prostration. If cooked with vinegar even, the injurious kinds are not so likely to produce irritation.

Poke root, known by botanists as the *Phytolacca Decandra* is a poisonous substance by which I once knew a whole family seriously affected, it being mistaken for horse radish, and grated in vinegar. The leaf when young is sometimes collected, and eaten for greens with impunity, but even when thus used these should be parboiled in the first water. The symptoms, and treatment of poisoning therefrom are very much the same as that spoken of in respect to poison mushrooms.

The sumach, or poison oak, is a small bush with spires of bunches of red seeds set closely together, and though valuable as an astringent in sore throat, is an acrid, narcotic poison. The milky juice exuding from the stem when broken will cause irritation of the skin in some persons. Children should be cautioned not to eat the seeds, and in case of any effect therefrom, a mild emetic followed by strong coffee should be administered.

The mercury-vine, a common creeper about old fences and trees, is often a source of irritation to the skin of farm laborers, and is said sometimes even to cause an irritation of the mucous membrane of the lungs. Unfortunately for the doctors, the people often presume that it is the material from which calomel is obtained, and of course infer that the latter must be a terrible evil. Some are much more susceptible to the poison of this plant than others, and the same person will escape at one time and suffer at another. When the first symp-

tom of an attack supervenes, a strong oak-bark tea may
be used upon the part, but.if the eruption extend, prepared chalk should be freely applied several times a
day and the part bathed once daily with any mild acid,
as vinegar, buttermilk or tomato juice.

If the inflammation is severe and painful, an occasional
poultice will be required.

We may conclude this whole subject of poisons with
two or three remarks of caution, worthy not only the
notice, but of the practice of all.

I.—Never have any poisonous article in the house,
without its having *poison* plainly marked upon it, and
then let it be put away beyond the reach of the smaller
portion of the family.

II.—All prescriptions of the physician of which you
do not know the component parts, should either be destroyed, when no longer needed, or labeled as articles
not to be used without advice.

III.—All animal-killing powders, or salves, should be
marked and cared for as poisons; for what will kill
small animals will often affect larger ones; and these
things, made palatable for the lower species, are sometimes relished by children, to their destruction.

IV.—Teach those under your control not to eat any
vegetable, leaf, or other article, not knowing its action.
While I write, a little patient lies in a neighbor's house,
killed by eating the ends from a few matches.

V.—All flowers with the cups turned downward, or
hooded, like the bachelor's button, and all stalks which
exude a milk-white juice when broken, are to be regarded as poisonous.

11*

VI.—All paints, whether of oils or water colors, should not be allowed to be placed in the mouth.

It is related of a celebrated English punster, that when told, just before his death, that he had swallowed some ink by mistake for the medicine the doctor had left at his bedstand, he instantly asked for a sheet of blotting-paper to take after it; and while such presence of mind is eminently desirable, where more noxious poisons have been employed, it is better still, by caution, to avoid these accidents, which are so frequently attended with serious consequences.

PRECAUTIONS AGAINST CONTAGION.

For all practical purposes, we may speak of contagion as of three kinds.

I.—That arising from specific diseases, as small pox, scarlet fever, and the like.

II.—That arising from marsh miasm, or vegetable decay, such as gives rise to chills and fever, bilious remittent fevers, &c., and

III.—That arising from effete or decaying animal substances, such as dead animals, or the exposed secretions of living ones, causing typhus fever, &c.

With cases of undoubted contagion from the first class, all are familiar. The effect of particular localities is not generally so extensively recognized by the public, but facts, with which all physicians are acquainted, clearly show that many diseases are dependent upon a

poison generated and produced by vegetable decay.
Fevers have frequently been known to be confined to
the vicinity of a creek or pond recently drained; places
once healthy have become quite unhealthy for a time,
by the cutting down of a large forest, and the exposure
of its surface to the direct rays of a summer sun, the
upturning of rich alluvial land in new countries is
always the source of sickness, and did space permit, we
might go on to show that the connection of disease with
changes and localities is as direct as that of cause and
effect in general. There are certain marshes in Italy,
over which if any one passes early in the morning, or at
evening, they are sure to contract an ague, while at
other times of the day there is but little risk.

There is another point peculiar as to the miasm from
vegetable decay. Its effects, even when for a time overcome, seem still to act upon the system, and there is a
tendency at periodic times to their recurrence. I have
known this to manifest itself even for five or six years,
and that after the person was entirely removed from the
locality.

The third source of contagion is less frequently recognized than the others, but nevertheless, is real. Low
forms of fever have often been traced to decaying matters, to foul sinks or privies, or to masses of putrid
human filth. Typhus fever in its worst forms, often
has such an origin, and the uncleansed holds of emi-
grant ships, and crowded tenement houses on land, not
unfrequently originate and propagate this source of disease.

There are three or four general rules that apply to all

contagions, which may be referred to as applicable in all
cases.

I.—Peculiar states of the system are favorable to the
contraction of contagious disease. What these peculiar
states are we cannot always determine, but those feeling
unwell are more liable to contract disease, unless labor-
ing under some severe affection. For instance, it has
been plainly shown, that an invalid from consumption,
cancer, or such like formidable disease, is not so liable
to skin affections, while persons with a recent bad headache and disordered stomach will more readily contract
those disorders.

II.—On an empty stomach we are more liable to disease than after eating. Food seems to support the system against contagion, and to occupy it on something
more important and agreeable. It should be a universal rule, never unnecessarily to expose yourself just before a meal-time.

III.—Morning and evening are more favorable to
most contagions than the middle portion of the day.
Dew seems to make some contagions more diffusive, the
secretions become more vitiated during the night, so as
to accumulate by morning; and at evening the tone of
our system is rather lower than at morning.

IV.—Cleanliness is an important preventive of contagion. Soiled clothing and foul skins are its hot-beds;
and where the skin is in proper condition, there is reason to believe it will rid the system of contagious influences, even where a slight amount of infection has
occurred.

V.—It is often important to avoid an infected district. Disease often seems to limit itself to special localities. All contagions are not equally diffusive; and just as you find different soils on opposite sides of the same road, or fogs resting thicker upon one portion than another of points in which we can discover no difference, so disease sometimes seems to choose its own margin lines, and if we cannot account for them, it is our business to respect them.

VI.—Fear is an undoubted source of contagion. When we see what an emotion of the mind can produce, as faintness in one, and a blushing cheek in another, and joy in a third, and so on, according to the character of the impression made, we need not wonder that fear should predispose a system to infection. Abundant are the instances I might mention where the frightened ones have contracted the feared disorder, and perished, while the bold and fearless have escaped. Our very fears are argument against our exposure, but where duty requires it, we may overcome this by having that love which casteth out fear. Very rarely have I known a devoted nurse who, from affection and without dread ministered to a friend, to contract a contagious malady.

VII.—Condiments or acrid substances often seem to protect from contagion. They occupy the mind, the stomach, and the mouth, and with these three busy, we are not so apt to be affected. Cloves, camomile, tobacco, or a cup of coffee, undoubtedly are of some service. Alcoholic stimulants are objectionable, because while they may protect for the moment, when their effect subsides we are more liable than ever to infection.

VIII.—Most contagions are more likely to be contracted by a direct inhaling of the breath of the sufferer, or by contact with the clothing or furniture of the room. In a foul locality I have often been particular not to sit down, and to wash the hands by pouring water upon them after my examination of the patient has been completed.

IX.—A very important method, perhaps the most so in preventing contagion, consists in the use of those articles known as absorbents. Burnt sugar, vinegar, and strong odors will conceal unpleasant effects, but the absorbents here alluded to actively neutralize them, and thus a much greater benefit is secured. Chloride of lime or chlorinated soda, known as disinfectant fluid, or Labarraque's solution, should be kept on hand in every house. The chloride of lime is easily affected by exposure, and is apt therefore to become changed, but the chlorinated soda-solution is easily kept, and either are cheap enough to be secured by all. A tablespoonful of either of these is worth more as a disinfectant, and deodorizer than a peck of sugar or a pint of cologne. In every disease communicable from a patient these should be freely used. A half tablespoonful of chloride of lime may be placed in a saucer, and two or three of these placed about the room once or twice a day, or the same quantity of the soda-solution may be used instead. A little in the water in which the attendant washes his hands, or around the person of the visitor, will be of service. Next to these powdered charcoal is valuable in the same way, and slightly moistened it will take up from the air many of its noxious qualities. Plaster of Paris is also a deodorizer and thus may with great pro-

priety be freely scattered about an infected locality.
Lime is valuable not so much as a neutralizer, but by
its active power, it in due time expels the noxious
effluvia. It is more valuable after the other articles
have been used, but even without them is of much service. The farmer may easily by the following process
prepare a chloride of lime which is excellent both as an
addition to manure heaps, and as a means of correcting
the sources, and the products of contagion. Dissolve
one bushel of salt in water, and pour it over three
bushels of lime fresh from the kiln. Stir it over, under
cover every other day for a fortnight, and you have a
mixture of chloride of lime, and carbonate of soda
which will answer very well for use about the out-
buildings.

X.—Great care should be taken to keep neat cellars,
and basements, and they who allow vegetable or animal
matter to decay therein, as is too often the case, or who
neglect frequent cleansings, and whitewashings often
suffer ailments from this source when their cause is entirely unsuspected. We have many times been surprized
at the carelessness of careful men in permitting turnip-
tops, cabbage, and other things to heat in the cellar,
and evolve most injurious gases, and we have much
sympathy with those who classify cleanliness among the
cardinal virtues. Vegetable or animal decay of any
kind should never be allowed to proceed in a confined
locality and masses of human excrement are especially
liable to give rise to serious disease.

XI.—The management of the sick room and of the
patient, has a very important bearing upon the lia-

bility to disease, but this will be noticed under that head.

The first class of contagious diseases are chiefly those affecting the skin. Children are much more liable to contract them than adults. The contagion is much more likely to be communicated after the eruption than before, inasmuch as this is the chief source of the contagion. Children should not be exposed to those affected, until after recovery is fully established. These diseases are not generally communicable at a great distance. Small pox, though very contagious, is easily expelled from clothing by a good airing. The contagion of scarlet fever is more difficult to expel from a house than that of any other of the specific contagions. The direction of the wind, as it respects the relative position to the patient, has an effect in conveying contagion. It has been asserted on good authority, that the infection of scarlet fever cannot be in a still atmosphere, transmitted more than five feet from the patient. Woolen clothing is much more likely to hold and transmit the odors which convey contagion, than other garments. Where a person has been exposed to the contagion of small pox, he should be forthwith vaccinated, and hearty food should be avoided for a few days. Where there has been exposure to scarlet fever, an occasional warm bath and the tincture of belladonna may be employed. In all these skin diseases, especial reference should be had to cleanliness, and to avoid exposure to cold, inasmuch as all of them are made much worse by any severe cold settling upon the lungs.

Where the contagion is dependent upon miasmatic influence, as in ague and the like, there are three special indications.

I.—To remove from, and avoid the vicinity of stagnant water, marshy lands, and places where for any cause large masses of matter are suddenly exposed to the heat of the sun.

II.—If unavoidably located in such localities, by draining, underdraining, cropping, and the use of plaster, lime, and chlorine, endeavor to get rid of the contamination of the air.

III.—Avoid sleeping in low apartments, or on the first floor, and avoid exposure to the dampness either of morning or evening.

A stagnant pond, a damp atmosphere, decaying cabbage leaves, potatoes, or other articles in the cellar, and water injured by the presence of some noxious substance, have often been the recognized and oftener still the secret cause of many an alarming case of sickness, extending through a whole family or neighborhood.

In the third class of morbid influences arising from decaying or effete animal matter, perfect ventilation, frequent washing of the patient, cleanliness in all its forms, and the free use of chlorine, plaster, charcoal and lime, form the best guarantee of escape from the wretched poison thus generated.

THE SICK ROOM.

The management of the sick room is an art, I had almost said a science, which has very much to do with the comfort of the patient and with the result of his disease. Worse than a poor doctor is a poor nurse, and, being a good one, is a very noble and difficult thing.

There are a thousand little attentions, and turns, and fixtures, and kindnesses, which, small though they may be, are of mighty consequence to the patient, and no one but the dependent invalid knows the distinctions between a good and poor attendant. I would scarcely venture to define what constitutes the former, but a few points may be specified, which must be attended to by all who would endeavor to deal faithfully by the sick.

I.—The importance of ventilation for the health we have already spoken of, but it is of still greater consequence to the diseased. Fresh, pure air, without any draught, is not the terrible thing that listed doors and stuffed key-holes might lead you to imagine. The sick room, the most of all others, needs, each day, a careful airing, and the raised windows, with the beams of the morning sun entering the chamber, or the mild refreshment of its setting rays, will be like health to the bones. Catching cold, to a sick patient, almost always results either from exposure to a draught of air, from an over close room, or from a very sudden change of its temperature, and is very rarely connected with the careful admission of air from without. A chimney and open fireplace, with a little fire, are very desirable, as providing a constant ventilation, and favoring the egress of the vitiated air.

II.—The regulation of the heat is another important matter. A thermometer is a very good companion on this score, and a temperature varying from 60° to 70° is usually desirable, although this may need to be modified to suit the condition of the patient. If a wood fire is used, care should be taken to guard against the extreme of heat caused by a sudden blaze, and the comparative coolness from dying embers. Smoke should always be carefully guarded against. If coal is employed, great pains must be taken that none of the offensive gas escape into the room, and a pan of hot water upon the stove will secure the requisite moisture.

III.—*Cleanliness* is often too much neglected in the sick room, even by those who desire to be neat. The fear of giving cold often deters from changings and washings, which would be attended to in health, and thus the patient suffers from obstructions to free exhalation. In cases of typhus fever in most of the hospitals it is the direction to have the patient bathed each day from head to foot, and although this might seem to be exhausting, and the necessary exposure to be dangerous, yet experience shows, that those thus treated are really refreshed by this process. There may be such a thing as too frequent disturbance, but it is a safe rule to bathe, and wash, and change the patient often enough to secure the most scrupulous neatness. Bathing a portion of the body at once, and drying it with a flannel or huck-a-back towel, and slight friction of the skin, are pretty good securities against cold.

The bed should be either of hair or straw, with a woolen blanket between it and the lower sheet, and woolen, more than cotton, should predominate in the upper covering. Those naturally of cold temperament, or who are suffering from chronic diseases in which much heat is not evolved, may with propriety be placed upon a feather bed, but this should be small, so as merely to ease the pressure of the patient's weight. The Dutch plan, of feathers all about, is not only abominable in health, but still more so in disease. The frequent changing of the patient from one bed to another is oftentimes very desirable, and the refreshment more than counterbalances the trouble; if carefully done by the use of a cot brought alongside of the bed there is need of little or no exertion on the part of the sufferer.

IV.—The exclusion of light is sometimes carried to too great an extent in the sick room. In diseases of the eye or of the brain, or wherever there is extreme sensibility, a dark room is desirable, but where these do not exist a moderate degree of light prevents gloom, is less conducive to the dreary wandering, which is sometimes increased by manufactured twilight, and besides, light as well as air is conducive to health. Windows hung with bed quilts, beds with canopy and curtains, doors with list, and all such like preventives of health, are, fortunately, passing away before more correct methods of rendering the sick room pleasant and healthy.

V.—Company is too often one of the unintended nuisances of a sick room. Some have a strange curiosity to see how badly a sick person can look, and even when they know they can be of no use, are frequent in their visits, and pressing in their desires to behold their sick friend. To many the mere presence of a friend, and still more, conversation, or the expression of their countenances is wearisome and exciting, and I have often been astonished at the injury done, even where, at the time, the call seemed acceptable. In the dreamy unconsciousness of evening each visitor has proved a ghost, whose shadow haunts the sleeping hour, and adds to the restlessness of the fevered brain. Frequently have I seen a patient suffer for hours from a visit or remark which at the time seemed to be a source of enjoyment. All are grateful for attention, but you will find it to be the almost universal will of those seriously sick, not to see any save their attendants, or those for whom they make special request. I would never have a friend denied the opportunity of a parting look, when it cannot possibly do harm, but where the physician or the nurse think it imprudent to invite you into the sick chamber, do not feel that you have been slighted; if you are a true friend, you would much rather deny yourself the pleasure, than to be the cause of injury to the sufferer. These minds of ours are wonderful constructions, and when the body is diseased it is strange how little circumstances may set them all ajar, and make each word and thought a goading thong to a frame already racked with pain. Inquiries at the door for those we love, or offers of readiness when assistance is required, are ever grateful tokens of a friends regard, but the same cannot be said of some more questionable attentions. Do not always conclude that the patient is very sick because the doctor prohibits company, for

perchance, he does it because he desires that he shall
not become so.

VI.—Cheerfulness, without levity, is always desirable
on the part of attendants; not that which exhibits itself
by an evident effort which the patient will detect, nor
by constant conversation, but rather the quiet expression of serenity, which studies every comfort, and is disposed to make the best of every symptom that a good
judgment will permit. Quietness, without moroseness;
sedateness, without sadness; a sensitive regard to each
want, without alarm or manifest anxiety; conversation,
without loquacity; and pleasantness, without lightness,
is the every-day rule for the nurse. Politeness has been
defined to be kindness kindly expressed, and the definition in its full is the climax of the art of attention.

Besides the due ventilation of the sick-room to which
we have already referred, too great care cannot be taken
to prevent the air from becoming contaminated. The
discharges from the patient, whether liquid or solid,
should be promptly removed from the room, and chloride of lime or chlorinated soda is never out of place in
the sick apartment. Not only is its absorbing and
neutralizing power a benefit to the patient, but conduces
to the health of the nurse, and of others who have reason
to frequent the house.

Such are the chief points in respect to the management of the sick-room, and every motive, both of faith-
fulness to the patient and of interest to ourselves, prompt
to a careful observance of these suggestions.

DOSES OF MEDICINE.

The amount proper to be given of any particular medicine is modified by a variety of circumstances. Age, sex, temperament, and the severity of the disease, all need to be considered as guides, and the remedy proportioned to the indications which these suggest. Women in general require smaller doses than men; persons of a nervous temperament are more readily affected by stimulants, and also by cathartics, than those of sanguine temperament; those who have frequently or habitually used opiates or stimulants, will bear increased quantities; and persons suffering intense pain, will require larger and more frequent doses than those slightly ailing. Various attempts have been made to furnish a schedule suited to all medicines and all ages, so that it might be known at once how much to give of every medicine to persons of any particular age, but it is apparent that no absolute rule can be made to apply to all cases. The following is a safe guide, and will never be too much, if any is required, while in severe pain more will be indicated. The proper dose for a person at twenty, or full age, being known, give to every person above ten an average dose thus: 11–20ths for eleven; 13–20ths for age of thirteen; 15–20ths for age of fifteen, and so on. For every person younger than ten, add their age to the ten and give the fractional part of the dose for twenty. Thus at nine, 9–19ths of the dose for the age of twenty; at eight, 8–18ths, or 4–9ths; at six, 6–16ths, or 3–8ths; at four years, 4–14ths, i. e., 2–7ths; at two years, 2–12ths, or 1–6th, and so down to ten months; below this down to five months, one-half the amount for ten months may be given, and younger than this down to one month, one-fourth.

The doses of any thing liquid are measured by drops, drams and ounces; and in order to be accurate as to them, it is only necessary to know what proportion they bear to the common vessels of liquid measure. With a small tin gill-measure and the following table, it is very easy to calculate the proportions:

  60 minims or drops make a dram.
  8 fluid drams make an ounce.
  4 fluid ounces make a gill.
  16 fluid gills make a pint.

According to the usual size of spoons, cups and tumblers at the present day, the table may be made still more practical by the following statement of proportions:

60 drops make a dram.

1 dram makes a teaspoonful.

4 teaspoonfuls make a tablespoonful, or half an ounce.

2 tablespoonfuls make an ounce.

8 tablespoonfuls make a gill.

8 fluid oz. or 2 gills make a coffee-cup or tumbler full.

6 fluid oz. make a teacupful.

All these up to the gill are not meant as even full, but full enough to be easily carried. By the use of the gill-measure, any one for themselves can by this table compare the size of various domestic measures, and determine any variation.

The articles are so few which it may be proper or necessary to be administered in a family, in the absence of a physician, that it will not be tedious to enumerate some of the most important, naming the doses for the ages of twenty, ten, and five years.

I.—*Cathartics.*—

		TWENTY YEARS.	TEN YEARS.	FIVE YEARS.
Castor oil..................	tablespoonfuls	2	1	½
Epsom salts......................	"	1½	⅔	½
Magnesia	"	1	½	¼
Rhubarb, powdered	teaspoonfuls	1	½	¼
Syrup or aromatic tincture.........	tablespoonfuls	1	½	¼
Sulphur	teaspoonfuls	1	½	¼
Cream of tartar...................	"	1	½	½
The two mixed equally...............	"	1	½	¼
Aloes	"	½	¼	⅛

Spirits of turpentine is an excellent cathartic with castor oil, and may be given as follows: At twenty, one tablespoonful of castor oil and one and one-half teaspoonful of turpentine, mixed; at ten, two-thirds teaspoonful of oil and one teaspoonful of turpentine; at five, one teaspoonful of oil and one half-teaspoonful of turpentine.

Senna, another common cathartic, may be prepared by pouring a half-pint of boiling water upon an ounce of senna, allowing it to simmer for half an hour, then strain it and put a little sugar and milk therein. At twenty, the dose would be four tablespoonfuls; at ten, two; and at five, one; repeat every three hours until it operates. Those who from feeble constitutions or debility are not of equal strength or development with the years of age, will of course need to be prescribed for according to their peculiar condition.

EMETICS.

These are sometimes quickly required in cases of poisoning, asthma or croup.

The doses of these should be repeated every twenty minutes until they cause vomiting.

DOSES AT		TWENTY YEARS.	TEN YEARS.	FIVE YEARS.
Ipecac powders	teaspoonful	½	¼	⅛
Mustard	"	2	1	½
Syr. Ipecac	tablespoonfuls	2	1	½
Hive syrup	teaspoonful	1	½	¼
Antimonial wine	"	1	½	¼

Emetics should always have given with them plenty of fluid, as warm water, milk or flax-seed tea.

As hive-syrup and wine of antimony are remedies of power, to children it is often best to give half the dose above named every ten minutes until vomiting ensues, instead of the full dose every twenty, and then discontinue it as soon as symptoms of vomiting occur.

SEDATIVES, NARCOTICS AND NERVINES.

These are medicines which have a tendency to quiet the system, lull pain, and produce sleep. If given too largely they are powerful for evil instead of good, and when a tendency to sleep has been established, they should always be discontinued.

DOSES AT		TWENTY YEARS.	TEN YEARS.	FIVE YEARS
Laudanum or tincture of opium	drops	25	12	6
Paregoric	teaspoonful	2	1	½
Magendies solution of morphine	drops	4	2	1
Tincture of Hops	teaspoonful	2	1	½
" Lavender	"	2	1	½
" Assafœtida	"	1	½	¼
" Valerian	"	1	½	¼

Laudanum should always be carefully corked. By evaporation of the alcohol it sometimes becomes thick and thus each drop equal to two or three of the regular preparation. Great allowance must be made for special cases in the use of opiates. While the weakly need less doses than those robust, where the amount given does not relieve the excessive pain in an hour, it may be repeated, and the friends of the patient should not rely upon their own knowledge only in cases of emergency, and soon send for one educated and experienced to detect the hidden causes of suffering.

Essence of peppermint and soda are often valuable aids to the action of narcotics where there is much pain. Regarding a half teaspoonful of peppermint, and one-sixth of soda as the dose for an adult you will easily compute by the rule above given the proper amount for a child.

The manifold other remedial agents which might here be named are usually not called for on an emergency, or fit for the handling of inexperienced persons, and we shall leave these to be prescribed entirely by the medical attendant. Where you cannot trust to the powers of nature, it is best to rely upon professional aid, unless the exigencies of the case are so great that delay itself is dangerous. In respect to medicines in general, we may remark that, like edge tools, they are safest in the hands of the well disciplined workman, and while we do not sympathize with those who claim the personal supervision of every ache and pain, we still feel the importance of having each one feel just how far they may safely go. As to the common nos-

trums and patent medicines of the day, there are several reasons why you should not be in the habit of using
them.

I.—In the case of hundreds of medicines celebrated
in the last century, when their composition has become
known, they have been found to contain ingredients
either perfectly harmless for good or evil, or else materials which should never be administered except under the inspection of a physician. One of our most
popular panaceas has in it a very respectable quantity
of corrosive sublimate, and many of those advertized
to contain neither calomel or quinine, have mercury or
Peruvian bark in some other form.

II.—They are usually prepared by persons whose sole
object is to make money, and who follow humbug as a
business. The recent detention and examination of let-
ters at the New York Post-office, showed most of these
medicine-men to be those who were engaged in various
plans for imposing upon the public.

III.—A physician educated for his profession, carefully acquainted with the different organs and functions
of the body, and knowing your constitution and the
peculiar state of the system, as no patent medicine can
know it, is more likely to benefit you than the chance
shot of an unknown remedy. An uncertain aim may
do for the flying bird or the bounding deer, but it is not
what we want when a human life is in jeopardy.

IV.—If you do not choose to have a physician, it is
safer for you to prescribe for yourself a medicine you
know something about, than to take an unknown article. Most men learn, or may learn, something of the

requirements of their constitutions by their own observation and experience, and sooner trust a knowledge of
your own case than the similarity of another's ailments
to yours, alleged by one who never saw you. The
poorest use to which a human stomach can be put, is
the making of it a juvenile apothecary's shop, and the
physician who studies carefully and thinks closely in
order to act, cannot but wonder at the rashness of those
who are ever ready to take everything that every body
recommends.

V.—Nothing is more deceptive or fallacious than the
common unprofessional testimony as to cures. Cases
are reported as precisely similar, and regarded as such
by the relations and by most observers, when the physician sees between them as much distinction as between
heart disease and apoplexy. Recoveries are claimed
where the future history shows merely a cessation of
disease, such as nature itself sometimes institutes.
Cures are boasted of where the malady has but been
concealed by placebos, and the patient, by hopeful sedatives, lulled to the tomb, and where real restoration has
taken place, three questions must be determined :

I.—How many others have been injured by the same ?

II.—How much had the medicine to do with recovery.

III.—How much sooner would it have been secured
by the care and skill of a thoroughly educated physician ?

For these reasons, and others which might easily be
named and illustrated, we regard the cases rare in which
common judgment will dictate the internal use of any
of these universal specifics, and it is better to provide

yourself from your druggist, with medicines whose properties are easily known, equally cheap and effective with those more dangerous and doubtful; or better still, to trust yourself to one educated in the science, and practiced in the art of prescribing.

ARTICLES TO BE KEPT IN EVERY FAMILY.

There are a few articles for which sickness may create a sudden demand, which should be kept in every family as it is often inconvenient or impossible to obtain them just at the time needed, and not only is the want far more than the worth, but serious consequences may result from the delay; the convenience of the physician as well as the patient requires that they shall be at hand. They may all be placed in a small box by themselves ready for any emergency. The list is not long; the articles, not expensive, and the trouble of procuring them at once much less than it may be hereafter:

An injection pipe, a half pint or eight ounces.

Adhesive plaster, . . quarter yard.

A piece of sponge.

Simple Ointment,. . . . one ounce.

Mustard, pure, ... one box.

Ground flax seed, . . a half pound.

In grain do. . . a quarter pound.

Blister ointment, . . . one ounce.

Sweet oil, ... a half pint.

Castile soap, . . . a quarter pound.

Prepared chalk,   .   .   a quarter   "
Cream of tartar,   .   .   a quarter   "
Chloride of Lime, in tight bottle, a half pound.
Camphor, ...   one ounce.
Laudanum, ...   one   "
Paregoric, ...   two ounces.
Two doz. bilious pills, prepared by your doctor
    or druggist.
Borax, ...   one ounce.
Alum, ...   one   "
Hive syrup,   .   .   .   two ounces.
Ipecac in powder, .   .   .   half ounce.
Magnesia, ...   one   "
Epsom salts, ...   quarter pound.
Sulphur, ...   "   "
Castor oil, ...   half pint.
Spirits turpentine,   .   .   "   "
Seidlitz powder,   .   .   .   one box.
Alcohol, ...   one pint.
Tincture valerian, .   .   .   two ounces.
Tannin, ...   twenty grains.
Sugar of lead,   .   .   .   "   "
Hartshorn or smelling salts,
    well corked,   .   .   .   a small bottle.
Peppermint,   .   .   .   one gill.
Anise seed, ...   half ounce.

And boneset, wormwood, hops, and catnep, either in the
house or garden. The entire expense of these will be
small, and no one can afford to be without them. All
bottles and boxes should be carefully labeled and kept
in a cool, dry place.

DIETETIC PREPARATIONS FOR THE SICK.

In the considerations of the subjects of food, drink, and diet, we have already alluded to the general considerations which are to guide us in the choice of articles of food under the various circumstances in which we may be placed. It may be well for us, however, here to add some remarks as to the modes of preparing the articles designated, and of adapting them to the desires of those concerned. No one who has, as a physician, noticed the different methods in which different families follow out the same directions, but that will recognize in this very thing many sources of diet errors to which the sick are exposed. For instance, one patient being permitted a little broiled ham for dinner, the one family will cut a thin and tender slice, carefully broil it by successive turning and dipping until it is tenderly done, without being raw on the one hand, or over-cooked upon the other; while another will furnish it in such state as to offend the delicate stomach. Even a potato, properly cooked, dry, and mealy, is a very different thing from a water-soaked root of the potato vine; and so by abundant instances, from actual observation, we might in cooking account for the reason of disagreement without attributing it to differences in the patients, or in the effects of the same food upon different persons. The stomach, after all, is not half so capricious as some imagine, and we can pretty generally, by studious attention, designate what will agree and what not, if the mode of preparation is as accurately defined and applied as the prescription of the article to be used. For all practical purposes we may, so far as food is concerned, divide diseases into three classes:

I.—Those in which no food is either desired or required.

II.—Those states of system in which there is an over-demand, and danger of repletion; and

III.—Those in which there is want of appetite, but an evident call for encouraging the desire for food, and a plain necessity for support. In inflammation or active acute disease of any kind, the stomach secretes very little gastric juice, and food, so far from supporting, lies undigested, irritating the whole system, and causing exhaustion instead of giving nourishment; and it is a fortunate provision that under such circumstances there is generally a loathing of food. In the second class of cases, as in the state of convalescence from acute disease, the general indication is to listen to the calls of the appetite; but there is need of judgment and regulation, lest an over-supply is consumed. For such there are two plans which can be pursued—the first to restrict to a definite quantity at specified intervals, and the second to allow a good share of such food as will appease the appetite, and yet is not very nutritious. In the third class of cases it is our business and art, as it were, to coax the stomach into digestion, and by nutritious niceties we often do as much in restoring the digestion as by less acceptable bitters and tonics.

In active fevers and inflammations, the great demand of the system is for fluids; and pure cold water, weak lemonade, flax-seed tea, barley-water, and the juice of water-melons or oranges, form a sufficient variety of

12*

drinks. Where the tendency is to a low grade of fever, or to rapid emaciation and debility, even before the acute symptoms subside we must convey to the stomach food which is absorbed without any great tax to the digestive organs, for they will not perform as in health. Here milk diet is most appropriate, consisting of cow's milk, arrow root, tapioca, corn-starch, and the like. In some cases, where it disagrees, even the cream will need to be removed from the milk; and if the bowels are much affected, even a large supply of starchy food is objectionable. Not unfrequently alcoholic stimulants are indicated in these cases, and even tea and coffee are of service. As soon as a step is made toward recovery, broth and meat may be carefully allowed.

In convalescence from nearly all diseases where the appetite is good, the sensible theory as to diet is to make choice of the most digestible of the ordinary varieties of food. Thus, if oil of any kind is desired, butter and cream should have the preference, to the exclusion of gravies and solid fats. Of meats, mutton, venison and beef boiled or roasted, just well done, should be chosen before fried hams, fritters, clams or any other of the everyday luxuries of working men. Raised bread a day or two old, should be provided instead of the warm. Of vegetables, preference should be given to those hereto-fore mentioned. Good ripe fruits should never be pushed aside by green gooseberries, currants, cucumbers or cabbage, although many of your friends may have eaten them all their lifetime with impunity; and milk, soft-boiled eggs and starchy food will usually be found to agree. Thus, the system must gradually be brought up to that state in which we would submerge all rules of diet into the one—to exercise, chew well, eat regularly, eat only enough, and then eat what agrees with you—the climax of all dietetic regulations.

Where there is absence of appetite, and yet evident call for support, as is often the case in chronic diseases, this unnatural state of things is usually owing to some serious disease, which needs both the care and the medicine of the scrutinizing physician; but it must be the aim of all to supply the natural stimulus of food. The most digestible kinds are to be preferred, as heretofore indicated; but condiments and flavorings and forms are the not to be despised methods of creating a relish and arousing the system to renewed activity.

We shall here subjoin a variety of receipts for the preparation of articles suitable for invalids, from which, according to the peculiar call of the case, and in conformity with directions given by the physician, selections can frequently be made.

Milk.—This is the food always proper and to be preferred where the indication is to give support to the system, and yet avoid anything stimulating or exciting. If taken fresh from the cow, it has the advantage of having the oily matter already in suspension, and is rather more easily digested. Where the stomach itself is much diseased, we have before remarked that it will often agree best where the cream rises and is removed. Where the powers of digestion are good, but there is emaciation from other diseases, the pure cream is often of much service. Milk boiled, by the addition of a little water to it, and the scum arising removed, we have often found to agree better than in its uncooked state. It is often spoken of as more binding, but only because it is more readily and completely digested. We have already noticed that bulk is necessary to digestion, and hence it is often important to use milk as an ingredient of other preparations, or rather to use them as a vehicle for it. So manifold are its uses in all departments of cookery, that we shall need to speak of it more specially in other connections.

Eggs.—These also are among the most nourishing and digestible of food, and those of the common domestic fowl are to be preferred. By the invalid they are to be eaten either raw or but moderately cooked. If used raw, a fresh one should be beaten up with a little wine or milk, as may be allowed, and a piece of stale bread eaten with it.

The oily matter of the yolk offends some stomachs, and the white or pure albumen is more digestible.

The best method of cooking eggs for the sick is to break them in a basin of boiling water, to which salt has been freely added. In breaking it should be held near the surface, and as soon as the white coagulates, it is done. Eggs boiled soft in the shell are next to be preferred, but fried they are unfit for the sick. A soft egg or a very hard one is more digestible than one cooked so as to be tough and dough-like. Custard is made by beating together three eggs and three tablespoonfuls of sugar, and then stirring them in a quart of milk, and cooking about fifteen minutes in a hot oven. This is only allowable where nutritious diet is indicated.

MEATS.

An English author defines common cookery to be a "device for rendering meats hard and indigestible." This is in actual practice too true, and to the invalid especially, becomes a very serious matter.

All meat is better not to be used very soon after killing. It should be carefully cleansed and hung in a cool airy place for a few hours. A chicken itself eating in the morning, and eaten at noon is not in the best state for delicate stomachs, and so with other meats. By keeping a short time they become more tender without undergoing any unpleasant change.

As a rule, boiled meats are the best for the invalid. Thus the fibre is made tender, and if cooked in a covered vessel, with just enough water to cover them, the juices are mostly retained. Two or three hours over a slow fire is sufficient for the more tender meats, but if continued too long the nutritive matter is boiled out, and the fibre made closer by shrinking. Among English physicians, boiled mutton has always the preference for the convalescent but with us wool is the great object, and the meat is not so uniformly tender. Lean beef which has been well fatted is sufficiently palatable and tender. Either, nicely boiled, are excellent for the invalid.

Roasting is the preparation next best to boiling. This needs to be conducted over a slow, steady fire, and the meat to be frequently basted so that it shall not be hardened on the surface and raw at the interior. Just well done is better than rare or over cooked. An excellent plan for preparing meat is to boil it until nearly done, and then finish its preparation by roasting. This may be accomplished by draining the water from the dinner-pot or by removing it into a dripping-pan and placing it in an oven.

When meat is properly *broiled* it is in much the same state as when roasted, and is both palatable, nutritious, and not indigestible.

TO BROIL MEATS.

Cut the meat into slices, and if need be, make more tender by pounding. Place the gridiron over a bed of hot live coals. Have at hand a small tin, say a gill, of warm water to which a little salt, pepper and butter, have been added. Dip the meat into this and place it upon the gridiron; when dry, dip it again into the mixture, and keep thus dipping and turning until it is well done without being scorched. Then put it on a plate and pour the rest of the warm juice from the basin over the meat. Even veal when well fatted is sometimes permissible prepared in this way. Fried meats are never proper for those of weak digestive powers. Pork in any form is seldom allowable in any condition of ill-health, and we are not much surprised at the opinion of Adam Smith, who declared fresh pork and tobacco as permitted inventions of the evil-one. There are, however, rare cases where salted pork, and properly prepared ham do not seem to disagree. Of these the interior portion is always to be preferred as it is not so much affected by the salt or smoke, and is partially preserved by the exclusion of air as well as by other means. Pork or beef corned for only a few days, and then boiled often agrees well with the stomach. Salt pork or ham boiled is better than prepared in any other way, but the invalid will always do well to avoid fresh pork or the fat of smoked meats in any form.

There are some species of wild game that are well suited to the dyspeptic or convalescent. Tender venison, boiled or broiled, is savory and highly nutritious. Squirrels or rabbits, without being so rich, are admirably suited for weak and delicate states of the system. Of birds, the smaller and white-fleshed are to be preferred. The domestic winged fowls are not so tender and digestible as most of the wild, but these, too, may be so prepared as to be well suited to weak digestive organs. The boiled or broiled half grown chicken is the best of these; boiled turkey next; but the aquatic species are rarely advisable for the invalid.

Stews, although partaking something of the nature of boiled meats, are somewhat different in their dgestibility. The meat may be allowed, but the vegetables, potatoes especially, are in the cooking so mingled with fatty matter as often to prove indigestible.

The various kinds of white-fleshed fish form excellent articles of food in the first stages of convalescence, while a little irritative fever still remains, and the propriety of allowing more hearty meats is doubtful. These are always to be preferred boiled or broiled, as directed for meats. The best sauce or dressing is fresh cream, slightly flavored with vanilla, cinnamon, or some other condiment.

The most digestible of the shell-fish is the oyster.

This, different from most animal food, is more digestible raw and well chewed, than when cooked. This fact modifies the extent to which they should be boiled when prepared in this way. To prepare an

Oyster Stew.—Take a half-gill of water, add to it a little butter, salt and pepper if needed or allowed, and let it boil in the sauce-pan; then add a half-dozen oysters, without juice. When it boils up again, pour in one-quarter of a gill of hot milk; stir and let it boil up once more, and the oyster stew is done. In boiling milk, you always add a little water to prevent burning.

The other varieties of shell-fish are of doubtful propriety for the sick, and lobsters, muscles, and stem-clams had better be left for those who consider themselves hardy perennials. The juice of the clam, roasted or boiled, is relished by some dyspeptic patients, and is allowable where a slight cathartic effect is desired.

From the more substantial meats we have broths, soups, and what is sometimes known as the essence of meat.

In preparing any of these, the design is to boil the meat in such a way as that the juices shall be extracted and mingled with the water in which it is placed, and this so flavored as to render it palatable. All of them, therefore, need a slow fire, and continuous boiling in a covered vessel, with just sufficient water to prevent scorching. Salt, a little pepper when allowed, parsley, and other condiments and vegetables may be added, to impart a pleasant flavor. A broth differs from a soup, in being less nutritious, but especially in all the oily matter being removed therefrom. The essence of meat differs from either, as greater effort is made to confine and secure all the nutrition of the meat, so as to convey to the stomach of the patient every thing it contains but the fibre itself.

We have heretofore stated that, as a rule, liquid is not so soon digrsted as solid food, because the water of the liquid must first be absorbed before the stomach will give attention to the solid matter combined with it. Yet, from this it would be hasty to conclude that for the invalid solids are always to be preferred. Tardiness of digestion is with them not always an evil, if, when it does begin, the food is in a soluble, finely-divided state, and easily assimilated. The greater bulk at first calls out rather a larger secretion of gastric juice, and there is not so much danger of over-supplying the demand for nutrition as when solid matters are eaten. Hence the law is not absolute, and the experience of particular cases will best inform us as to whether slops or solid food agree best.

A broth or meat tea is prepared as follows: Put the half pound of meat to be used in a pint of cold water, to which salt and a little parsley or thyme have been added. Let it boil slowly for a couple of hours, adding water, if necessary, to prevent burning. The utensil used should be of small diameter, and kept covered. Any oily matter or scum rising to the surface should be carefully skimmed off every half hour. A half tablespoonful of rice, and a slice of onion, carrot, or parsnep, may be added a half hour before the broth is done, if their flavor is desired, but the pieces should be removed.

You thus have a preparation well suited for those who are just beginning to need animal food.

Meat Soup.—To a pound of meat, with a marrow-bone in it, add a quart of cold water and a little salt, and place it over a steady fire. After boiling an hour, add a little celery or parsley, some pieces of dry bread, and a few slices of such vegetables as are agreeable. Then boil an hour and a half in a close vessel, stirring occasionally to prevent scorching. If rice is used, as is best, it should be added an half hour before the soup is done, as, if put in too soon, it becomes sticky, and is liable to burn. In some cases, the pieces of vegetables had better be removed from the soup, as they may occasion griping and indigestion.

Essence of Meat.—Cut a half pound of lean but tender beef into strips after it has been well pounded, and place it in a bottle which will hold twice as much. Add a little salt and flavoring, and place the bottle, after being corked, in a pot of water, just warm, and boil for two or three hours. Then strain the juice through a seive or muslin, pressing all the liquid substance out from the shreds of meat, and you thus have a very nourishing preparation.

VEGETABLES.

But a few words need be said of these. The invalid or convalescent should be mostly restricted to potatoes, carrots, parsnips, tomatoes, and asparagus. A potato, to be eaten, needs to be well done, flaky, and easily broken apart. New potatoes, cutting like a solid rind, are not fit to be used. An amateur, as well as practical agriculturist and feeder, directs that, "after being washed, and soaked for a quarter of an hour, potatoes be put in a sauce-pan of cold water, sufficient to cover them, and, when this boils, let another cupfull of cold water be added, and then let them boil again until done." When soft, which may be known by trying them with a fork, pour off the water, and let the pot of potatoes, with the cover off, stand four or five minutes over a gentle fire until the remaining moisture evaporate. This last we regard as a very important direction, but, by attending to this only, and putting them peeled into hot water, with a little salt, we have usually found them mealy.

The *sweet potato* should be cooked by the steam of the water, and this may be done by placing a few sticks over the bottom of the pot, and then it should be dried out by the same process as the other. Either the common or sweet potato are palatable and quite digestible when roasted. After washing, a small piece should be cut from each end, and the potato placed in an oven not at first very hot, but sufficient, before they are quite done, to brown the skin. Both kinds of potato are rendered watery by being overdone.

The sweet potato should at first be eaten moderately by the invalid, as it does not digest well with all, while to some it is found highly nutritious. A good mealy common potato, well masticated, is among the first vegetables to be allowed the convalescent.

Tomatoes are less apt to offend the stomach cooked than raw. For the invalid, they should be prepared without adding any thing but salt, and a little sugar when just done. They should be cooked half an hour in a covered saucepan, and should only be just ripe. At the first use, the seeds may be removed, and only the more solid parts boiled.

Carrot and *parsnep*, after being well scraped and cut in slices or round pieces, should be placed in a saucepan of lukewarm water sufficient to cover them, and cooked until very tender. After a couple of hours, or before, if done, pour off the water and add a little seasoning and butter, if allowed. If preferred, when about half done, they may be removed, and broiled slightly brown upon the gridiron.

Asparagus needs boiling about half an hour, and for the invalid should be dressed with flavored cream.

The question is often asked as to the digestibility of the *cabbage* tribe. Where there is a tendency to flatulency or weakness of the stomach from exhaustion of its powers, cabbage, cauliflower, and the like are objectionable, but properly prepared they do not offend as much as is supposed. Raw cabbage, cut up with vinegar and well chewed, digests more rapidly than when boiled with fat pork, but this does not prove it when alone more digestible raw than cooked. In the one instance its digestion is hastened by an acid, and in the other retarded by grease. If permitted by the physician, it should be boiled in water well salted, with a small piece of fresh beef.

There are two or three kinds of vegetable soup or broth which may be allowed. A little celery or parsley, boiled for an hour in a quart of water, to which an onion, potato, and sliced carrot have been added, makes a pleasant liquid, to be eaten with toast bread.

A gill of the dry white bean, boiled for three hours in a pint of water, to which salt has been added, makes a soup often relished by the recovering. The vessel used in cooking must be of small diameter, and well covered. A little cream added to the soup just after it is strained, and while hot stirred through it, much improves the flavor.

Green vegetables in no form can be recommended to the sick; and although green peas, corn, beans, and cucumbers may do for the working healthy farmer, they affect differently the dyspeptic or weak invalid. The asparagus already mentioned, and, where there is constipation, the rhubarb or pie-plant, well stewed and sweetened with loaf sugar, may be admissible, but beside we know of no other.

FRUITS.

Of these the ripe, healthy peach, the full-grown, sweet-cultivated blackberry, the large, ripe strawberry, and roasted apples, are best for the invalid. Pears, plums, grapes, currants, and melons must be eaten cautiously, if at all. Our own prejudices incline us to regard ripe fruits as meant to be eaten and as good for everybody; and as they are a class of aliments different from meats or grains, much reliance is to be placed upon the experience of each particular individual. To some the seeds are irritating, to others promotive of digestion. With some one fruit will act as an astringent, and to another the same as a cathartic, and hence much depends upon particular indications.

Of stewed fruits, well peeled, healthy, fresh, dried

peaches, apples, and prunes are the best. They need to be stewed over a slow fire, with water just sufficient to cover them; and when fully tender, add just sufficient sugar to sweeten them, as, if too sweet even when eaten, as they should be, cold, this is apt to cause acid eructations.

DRINKS.

Of many of these we have already spoken, and need here only describe the mode of preparation for the sick.

Cocoa.—Boil an ounce of the prepared cocoa-shell in a half-pint of water, then add a half-pint of milk, and let it simmer over a slow fire for an hour. Sweetened with sugar it is ready for use.

Coffee.—Put a tablespoonful of fresh-roasted fresh-ground coffee, with which the white of an egg or a little isinglass has been mixed, to a pint of cold water, and let it boil up and then simmer for an hour. Then strain and add half the quantity of boiled milk, and sweeten to suit the taste. If too strong, it may be then weakened with boiling water.

Black Tea.—Put a teaspoonful to a half-cup of boiling water and let it boil up under cover for three or four minutes, and then add water, milk and sugar to make it of agreeable strength and taste.

Barley Water.—Put an ounce of best barley to one pint of water. Add lemon peel, a little cinnamon, sugar, and a few raisins. Boil down to suit the taste and drink when cold.

Tamarind Water.—This is made by pouring a pint of cold water upon an ounce of tamarinds, stirring it well and letting it stand for an hour. Then strain, and sweeten a little if desired.

Toast Water.—This is prepared by pouring boiling water on baker's bread toasted nicely brown, and when cold it should be strained and put in a tight bottle.

Rice Water.—To two tablespoonfuls of rice add a quart of water; boil for two hours, and continue, if need be, adding water enough to prevent it from becoming thick. Then add sugar, cinnamon, or other flavoring; strain by pressing through a linen rag, and use when cold.

Flax-Seed Tea.—Take of the fresh seed unground, two tablespoonfuls, add one pint of boiling water, and let it simmer for an half-hour. Then strain as before and add freely of lemon, nutmeg, allspice, or other flavor. If for colds, it should be used hot; if for fevers or bowel affections, wait until cold.

Wine Whey.—Boil a pint of milk, and add to it while boiling, a gill of Madeira or sherry wine. Let it boil again, and after standing a few minutes strain off the whey from the curd, and sweeten to the taste.

JELLIES.

These are prepared by boiling the juice of the article used with sugar, so that the two unite to form a compound consisting mostly of sugar and gelatine. Many of these are nutritious, and in small quantities easily digested, and both palatable and permissible for the sick. A selection of a few preparations will afford a sufficient variety.

Quince, Raspberry, Blackberry, or any fruit jelly,

is made by squeezing out the juice, adding a pound of best sugar to a pint of juice and boiling it down to a proper consistence, usually for fifteen or twenty minutes, over a slow fire. The fruit needs to be only just ripe, as its state has much to do with the rapidity and ease of making the jelly. Where it is desired to make apple, pear or quince-jelly, it is most expeditious to boil the fruit just done in very little water, and then strain the juice through a flannel bag, and boil down as before. The blackberry, apple and quince-jelly are the best for the sick, although the more acid seem frequently to agree.

Arrow-Root Jelly.—Mix three tablespoonfuls of the best arrow-root thoroughly in a teacup of water. Boil a half-pint of milk, and when just boiling, stir in it the cup of arrow-root, and enough sugar and flavoring to impart a pleasant taste, and then let it boil again for a few minutes, until thick. It may be eaten hot or cold, and milk, wine or nutmeg added, if the doctor will permit.

Tapioca Jelly.—Put three full teaspoonfuls of tapioca to a pint of cold water, and change the water after two hours; then let it stand for an hour; then add a tablespoonful of sugar and boil gently for an hour, or until it has a jelly-like appearance. Sago may be prepared in the same way, and either eaten with such additions as will render them palatable.

Rice Jelly.—Boil a half-teacupful of rice in sufficient water to cover it until just soft, then add as much sugar and a little more water, and boil until it becomes a jelly-like mass. Raisins or a little milk may be added just before done, but neither should be eaten if the stomach is weak.

Isinglass Jelly.—This is not, as some suppose, made of the mica used for stove doors, but is itself a purified jelly, prepared from fish. To fix it for the invalid, boil one ounce in a pint of water slowly down to one-half, then add a half-pint of boiling milk, and sugar, flavoring, &c.

Blanc Mange.—Break up three sheets, or about one ounce and a-half of the best isinglass into a quart of milk. Simmer for half an hour over the fire. Then add a cupful of sugar, flavoring it to suit, and strain through a cloth into forms, and set aside until cold. When done, the isinglass, except a slight scum on the surface, is mingled through the milk, and by using a greater or less quantity, its thickness may be regulated.

PREPARATIONS FROM GROUND GRAINS.

Of these, the chief in common use are wheat, Indian, rye and buckwheat.

Fresh bread of any kind is not admissible for the invalid, and, with its tough crust and sticky interior, is far worse than warm cakes prepared from the same materials. In the process of rising or fermentation, changes take place which render some of the materials less digestible than in their natural state, and as a rule, dough of any kind raised as by the common yeast, is not easily digested when fresh-baked. Bread, biscuit and cake may, therefore, be spoken of as fermented and unfermented; the first including all those preparations of flour in which leaven is employed in some form or other, and in which the rising, is dependent upon vinous fermentation, and the second, those in which the lightness is dependent upon gaseous or volatile matter set free in the dough, without this fermentation. As a rule, the latter warm, may be regarded as more digestible than the former when in the same state, but when both are cold, there is not so much difference. Many dyspeptics, however, have thought, from their own experience, that unfermented flour preparations agree with them better. The following is a good and familiar recipe for preparing unfermented bread :

Flour .   .   .   .   . one pound.
Carbonate of soda   .   . 40 grains.
Cold water .   .   .   . half a pint or more.
Muriated acid of the shops 50 drops.
Powdered white sugar . a teaspoonful.

Mix the soda and sugar thoroughly through the flour. Then add the acid to the water, and, with a wooden spoon, gradually mix sufficient into the meal to make a dough. Divide into two loaves, and put in a quick oven immediately. Biscuit may be prepared by intimately mixing a teaspoonful of powdered soda, and two teaspoonfuls of cream-of-tartar, through a quart of flour, and then adding sufficient unskimmed milk to make a soft dough, which will usually be less than a pint. Divide into small biscuit, and bake quickly in a hot oven. These, when cold or a little stale, are for most, more digestible than fermented bread.

Wheat and Indian cakes.—Take a teacupful of wheat flour and Indian meal each, and sufficient milk, just sour, with which to make a batter. Then add the white of one fresh egg and a half teaspoonful of soda, and bake on a hot soapstone or lightly greased griddle. If the milk is quite sour, sweet milk may be chiefly used. Equal parts of sweet milk and fresh butter-milk will answer every purpose.

Panada.—Place two or three soda biscuit in a saucepan, and cover them with boiling water. Boil up until the mass becomes pulpy, then strain off most of the water, and beat up the rest to the consistency of gruel, adding sugar, milk, or wine, if permitted.

Thickened milk, or poor man's living.—Boil a pint of milk, with a half teaspoonful of salt and a tablespoonful of water in it, then make a thin batter of three tablespoonfuls of wheat flour, and stir it into the milk. Let it boil up, and then take it off and empty it into a dish, allowing it to stand until cold. It may be eaten with a little sugar, milk, or wine, if necessary.

Indian Gruel.—Take three tablespoonfuls of best well-sifted Indian meal, and mix it into a smooth batter, with a cupful of water, then stir it gradually into a quart of boiling water, to which a little salt has been added, and cook, with frequent stirring, for fifteen minutes, or longer. If desired richer, milk may be added to the water, and, when done, sugar and flavoring employed. Eaten warm, it often aids the action of cathartic medicine.

Indian Cakes.—First make a gruel by stirring a tablespoonful of Indian meal in boiling milk, a little water and salt having been added to it. Then add a half teaspoonful of wheat flour, and bake upon a griddle. The advantage of this method is, that the Indian becomes thoroughly cooked, which is not the case in its usual preparation, and is much less likely to offend than when merely baked.

Mush.—Boil the water with a little salt, and prepare precisely as directed for Indian gruel, except that a third more of meal should be added, and the boiling continued, at least, twice as long. Indian is generally eaten without sufficient cooking.

The modes of preparing rye flour are so similar to those of wheat that it is scarcely necessary to notice any distinctions. It sours more readily than wheat, and is less nutritious. It is much the best in an unfermented state, and may be made into cakes, mush, &c. Rye mush is often beneficial in constipation.

Buckwheat is allowable to the dyspeptic only in the form of cakes or puddings, and often not even in these. When eaten fresh, the objections urged against fermented preparations do not apply with full force to it, and when made light, and baked with but little grease, they agree well with some. The best form in which it can be used by the invalid as a hot cake is as follows:

Make a thin gruel of Indian meal as before directed, then add to it an equal quantity each of buckwheat and fine unbolted wheat flour, stir the whole into a thick batter, and when cold add yeast in proportion of a large spoonful to every pint of meal used. Put away to rise, and when light, thin, if need be with water, add a little soda, and bake upon a hot griddle. This makes a cake much more digestible to many than the pure buckwheat.

PUDDINGS.

These may be prepared in various ways from articles already mentioned, but many of them are of very questionable propriety for the invalid. It is well remarked by an American author that "it is doubtful whether there is any way of boiling dough to make it fit for food, and generally speaking, puddings for the invalid are best made from rice, sago, tapioca and the like." Of boiled flour puddings the least objectionable are the Indian, the stale bread and the buckwheat pudding.

Indian pudding.—Take one quart of milk, one small cup of wheat flour, one half teaspoonful of salt, one teaspoonful of soda and a few raisins, thicken with Indian meal to the consistency of soft gingerbread, and bake for two hours. To be eaten with butter and sugar, molasses or any other sauce allowed.

Bread pudding.—Break up finely a pound of stale bread, pour over it a pint of boiled milk, and let it stand for an hour in a covered basin; then beat thoroughly two eggs; add a little salt and a very little soda; put it in a bag, and this in a basin, and place it in boiling water for a half hour. It may be eaten with sugar, wine or other dressing.

Buckwheat or hasty pudding.—Mix about four tablespoonfuls of buckwheat meal smoothly with a little water and pour it in a pint of boiling milk, to which a little salt has been added. Stir and let it cook for a few minutes. It should be eaten hot with a little milk, butter or molasses.

Of baked puddings, the best are the rice and the sago or tapioca.

Rice pudding.—Put two tablespoonfuls of rice and two of sugar to a quart of good milk, and bake for an hour in a hot oven. A well beaten egg and a little cinnamon or nutmeg for flavoring may be put with it if desired.

Tapioca pudding.—Soak an ounce of tapioca in a pint of cold water for an hour, then pour off the water and add a pint and a half of good milk, one well beaten egg, a tablespoonful of sugar and cook until the tapioca is well mingled with the milk. If a thick pudding is desired, a little flour also may be stirred into it.

CAKE.

In general, cake or pastry of any kind is not allowable for those whose stomachs are weak or much diseased, but there are two or three of the plainer varieties which may be allowed.

Of fermented cake the *rusk* is the most admissible. Take over night one cup full of yeast and sugar each, one-half cup of butter, two eggs well beaten, and two cups of milk, and mix into a batter. In the morning add sufficient flour to make a soft dough, and mould into proper size. Eaten cold they often will not disagree.

Of unfermented cake, the sugar-cake and molasses-cake, made plainly, are the best; and domestic cookery is so familiar with the mode of preparing them, that we need not describe it.

There are various other preparations sometimes prescribed for the invalid, but it is not necessary for us to enter into minute description here. Farina, maizena, and corn-starch partake of the nature of the substances from which they are prepared.

Tous-les-mois is a pure form of arrow root. Macaroni and vermicelli are the gluten or pasty matter of wheat, and are to be judged of by the nature of the grain from which they are prepared.

Sufficient variety has already been afforded from which selections may be made; and with the general principles enunciated in the articles on food and diet, the invalid will be able, with the direction of his physician, to make such selections as will please the palate, agree with the stomach, and conduce to that perfect digestion which is the *sine qua non* of perfect health.

POULTICES.

In the treatment of both external and internal inflammations, a poultice is often a powerful auxiliary for good, and as these are usually prepared by the patient or his friends, it is important for all to understand their object, their varieties, and the end sought to be attained in their use. The experience of most physicians will agree with that of Abernethy, who, when professor in the Royal College, declared them "either blessings or curses as they are well or ill made, and more commonly as they are made only irritating instead of doing good." The two great points aimed at in a poultice, as applied to an inflamed part, are heat and moisture. When a part is just beginning to grow red or inflame, cold ap-

plications are often indicated, and hence, you will remember, in speaking of bruises, etc., we directed cold
appliances to be used at first, but after a part is really
inflamed, and there is an over-supply of blood and a
stagnation thereof in the part, heat and moisture often
have the effect of so altering the condition of the tissue,
as to favor a subsidence of the inflammatory action.
Cold, as long as it is agreeable and the pain is not
severe—leeches. if the pain and redness increase, and
then warm applications we regard as the rule in cases
of external inflammation. Cold poultices are sometimes
spoken of, and scraped potatoes, turnips, apples, and the
like, popularly used in this way. Where cold is indicated, cold and moisture are both needed; but while
cold articles like these do very well when the applications cannot be frequently renewed, in general, cold
water or cold spirits, either of which evaporate more
rapidly than when mingled with solids, are to be preferred, because the rapidity of evaporation has much to
do with the degree of cold. If any of these are allowed
to remain to become warm, they are changed from the
design of their use. We shall, therefore, speak only of
warm applications under the term *poultice*, and the
starting point of these, as we have said, is heat and
moisture combined. Heat will not do alone, for this in
fact is apt to increase the pain of the part, and moisture
alone is not sufficient, for if so, cold water or oil would
answer as well, but it is the combination of the two,
which so softens the part as that it may distend, and the
blood, so to speak, finds room to get away from its stagnant position, or if it has become so changed as to ren-

der this impossible, it will thus be facilitated in breaking down into matter and getting away from the system. The first great point, then, in the preparation of a poultice, is to make it in such a way as the heat and moisture will be longest retained. Hence, in general, articles that will retain warmth the longest are best for poultices, and it is familiar to all that some things, both of liquids and solids, such as boiling milk, oil, and hot mush, cool much more slowly than others.

But there are also other designs to be accomplished. Air is generally injurious to a raw sore, and oil excludes it and is softening even when the skin is not broken, and hence oily matter or mucilage in a poultice are of no small value. Sometimes there is so much pain that we need to combine an opiate with the poultice, or there is an unpleasant odor which it is desirable to correct. For all practical purposes we may divide poultices into five varieties.

1st. The common or plain poultice.

2d. The oily poultice.

3d. The anodyne or quieting poultice.

4th. The antiseptic poultice, to correct smell.

5th. The stimulating poultice.

Of the first, heat and moisture are the sole objects, and of all the rest this is the starting point, but other qualities also are added thereto.

The first point then, is how to make a poultice which shall combine heat and moisture. The design is to apply it to the part moist and warm as can be endured. Take a basin of size just sufficient for your purpose, and scald it out with boiling water as you would the tea-pot to make tea, then put in it the ingredients you design to use; if bread, let it be finely crumbled. Then pour boiling water upon it, as much as it will soak up, and stir it until the whole is well mixed and softened. It should be about the consistency of soft-bread mush, and should be applied to the part for which it is intended as hot as it can be borne. After an hour, if it gets partially cool, a little very warm water may be poured over it, and this, once or twice will answer to renew it without removing it. The poultice, if spread over a large surface, should still be pretty thick upon the cloth, so as to retain the moisture. It should not be left on too long without removing it, for then it will do harm rather than good. No specific rule of time can be given, but as soon as the heat and moisture are gone it should be renewed. Pastes of any kind, as of wheat flour, starch, gum, &c., will retain heat much longer than dry substances. The common bread and milk poultice, or the bran poultice, prepared like the flax-seed, are among the best of this variety.

The oily poultice differs from the common only in the material of which it is made, and in having oil in some form or other incorporated with it. Oil retains heat, favors moisture, softens the skin, keeps the poultice from adhering to the skin, and excludes the air.

Mucilage as that of gum Arabic, or slippery elm is very much of the same nature and may be classified with these. Butter, lard or sweet oil smeared on the common poultice will accomplish some of the designs above named, but the ground flax-seed and slippery elm poultice are the two standards of this variety. In preparing the flax-seed poultice, the cup or basin 'of proper size should be scalded out, the meal placed in, and boiling water poured upon it. It should be quickly beaten like eggs until quite ropy, and is then ready for use. The slippery elm is prepared in much the same way. Care must be taken that the flax-seed meal is sweet inasmuch as it sours by keeping like other meals.

The Indian meal poultice contains some oil, and when used should be made just as thin bread mush, but it is hardly as light and pleasant as either of the others.

The anodyne poultice is prepared just in the same way as those above mentioned, except that an anodyne is added thereto or substituted for some one of the component parts. Thus in preparing the bread or flax-seed or elm poultice, strong boiling hop or poppy leaf tea may be used in mixing instead of water, or besides this a teaspoonful of laudanum, or tincture of belladonna or aconite, or chloroform may be added thereto, or bathed over the part just before the poultice is applied, and you thus have a soothing poultice.

The antiseptic poultice in addition to the primal object seeks to correct foetor and is therefore only needed upon an open sore where there is unpleasant odor. A half tablespoonful of pulverized charcoal, or a half teaspoonful of chloride of lime dissolved in water, or mingled with the poultice when just ready for use, will answer this purpose. Chlorinated soda, raw carrots finely scraped, or brewers' yeast mingled with the poultice have the same general effect.

The stimulating poultice is designed to expedite the separation of the dead flesh, slough or hold-fast, as it is variously called, or so to act upon the skin as to hasten the exit of matter beneath it. Often in these cases the application of some caustic or acid, or wash, directly to the part just before the poultice is applied will answer this purpose the best; thus, if the slough is already apparent, a little turpentine, or turpentine and syrup of Peru mixed, poured upon it, will be an admirable stimulant, or a little caustic, or burnt alum will hasten the breaking down. If not already a raw surface, soap and sugar applied under the poultice, or soot, or gun powder will hasten the opening.

A flax-seed poultice, made with boiling molasses or honey, instead of water, is often of much service. A slight portion of salt in a slippery elm poultice, if there is no raw surface, or if there is a teaspoonful of turpentine mingled through it, if the slough is large and tenacious, will aid in its separation.

A slippery elm poultice, made with boiling sassafras tea, moderately strong, is more stimulating than as usually prepared. The brewers' yeast, or even the bakers' common yeast, is also valuable to mingle with the poultice for slight stimulation. When these are used, the poultice should be prepared as the common or oily ones are, except more stiffly, and then the yeast stirred through just as you are about to apply it, will make it sufficiently thin. If put in at first, just as in hot mush, as the housewife terms it, "the rising is killed," and part of the value lost. A gill of yeast is about sufficient for a quarter-pound poultice.

Boiled beans or peas retain heat well, and made soft, are slightly stimulating.

These, we believe, embrace all the important varieties of poultice. In common you will hear many other varieties named, and peculiar virtues attached to them. With many, strange ideas exist as to this class of remedies. They seem to imagine that there is some attraction about them, like that of a magnet, or some pulling power, or some ability to draw matter right up into them, but this is only indirectly true of any poultice. Heat and moisture, and oil, soften so as to aid nature in finding a vent for pent up material; or, a stimulating poultice may help to eat away the hold-fast, or break up the tenacity of the tissue above the seat of disease, and thus hasten recovery. In this sense only is any poultice drawing, and provided with the material best adapted to accomplish these points, you will not, with this understanding, be led to attach particular virtues to certain wondrous cure-alls.

There is a class of appliances holding a middle position between poultices and fomentations, which are often useful, especially in internal inflammation. As in these cases there is often great pain, and anything of much weight is very uncomfortable, these must be made of the lightest materials.

Bran or oats heated upon a stove or in a spider until quite warm, and then moistened only by just as much hot water as it will readily absorb, and placed in a bag of proper size, furnishes a most excellent application when one of such kind is required. The bran should be prepared fresh, and often changed, as it quickly becomes soured.

Hops warmed in the same way, and sprinkled with

warm vinegar or water, answer a similar purpose. They dry more quickly, but are somewhat anodyne. Starch or rice-grain may be used if preferred. Warm turpentine is often combined with any of these with advantage.

Thus, with definite ideas as to the aim of a poultice, we believe that many an erroneous practice in respect to them may be avoided. Every physician will testify how frequently, when he has prescribed poultices, he has called and found them, by improper preparation or attention, doing far more injury than good; and as this is not confined to the ignorant or careless, there is need of a careful understanding of the subject by unprofessional persons.

FOMENTATIONS.

These are of the same general character as poultices, having for their object to sweat the part and render it soft and pliable, so that, if possible, the inflammatory action may cease, and if not, that it may the sooner terminate in suppuration.

They are light, and less cumbersome than poultices, can be more readily applied, and leave no crumbs or hard morsels when they become dry. On the other hand, they need to be more frequently changed; if care is not used, the clothes about them will become wet, and thus the patient be much more liable to contract cold. In internal inflammations, and where care is used, they are sometimes even more valuable than poultices. As minute direction is necessary in respect to them, we will embody a description of their mode of use in one or two examples.

If, for instance, one of the lower limbs is to be fomented, the patient lying in bed, under him should be first placed a blanket or thick sheet, so that afterward any wet portion can easily be removed, then two pieces of flannel should be procured, one should be placed in water nearly boiling, and after it has lain long enough to become thoroughly heated and soaked, it should be wrung as dry as possible and quickly wrapped loosely about the painful surface. If a piece of oiled silk can be had, this should be wrapped about over it, as thus the heat and moisture are longer retained; or if not this, a dry cloth should be placed over it, so as not to mess the upper sheet. This will probably keep sufficiently warm for twenty minutes or half an hour, when the other piece of flannel, being already prepared in the same way, this should be removed and the second applied quickly, as before. Thus you may keep up the fomentation for any length of time required, and it will often be found exceedingly grateful to the patient. When done, every thing wet should be removed, and a dry flannel or piece of oil-silk applied to the part. Where the abdomen is to be fomented, though the part cannot so easily be wrapped up, yet a cloth easily removed should be placed beneath, and the same directions followed, even more care being taken to avoid exposure and to remove every thing soiled. There are other ways in which the idea of a fomentation may be carried out with benefit, where the patient is able to sit up. If it be the forearm, this may be placed in a kind of tin basin or stew-pan, and this partly filled with hot water, so as to cover the part, the tem perature being kept up by the addition of more from time to time. Where as in rheumatism, or lumbago, or severe cold affecting the joints, it is desirable to fo- ment a large surface, procure a tub of boiling water, in which a few ears of boiling hot corn or of herbs which will keep up the steam have been added, and set in it a stool ; then let the patient be covered with a woolen blanket, and sit upon it the tub and patient, both being surrounded with blankets so that the vapor cannot es- cape, and thus the pores of the skin will be relaxed. It should not be continued too long, and the subject should not go suddenly into a cold bed or cold room, but have these slightly warmed for his reception. If preferred, hot water in a small basin may be used sitting under a chair, instead of the tub, just enough alcohol being added to make it burn. A torch being applied, a slight flame will cause evaporation and perspiration. Rum, which is alcohol already diluted, may be used in the same way, a little care being taken that the flame does not catch to the woolen blanket, which it is not likely to do. You thus get a rum sweat, which in many cases will be of service. A small and cheap tin water-holder, made so as to fit over the abdomen or any rounding sur- face, is very convenient as a means of fomentation. It has a small stopper, by means of which the hot fluid can be poured into it, and thus the heat is applied and moisture occasioned by the difference in temperature, so that you obtain the value of the foment without the least mess.

As to the various kinds of fomentations to be used, they admit of very much the same classification as the liquids of which poultices are formed. Pure hot water is most usual and one of the best. If great pain, hops or poppy-leaf tea may be used, or laudanum or any other anodyne may be added to the water. Spirits of turpentine, chlorinated soda, spirits of camphor and the like may be added where special symptoms call for them, or any of the other liquids named in connection with poultices. Thus properly managed, fomentations prove valuable, not only in inflammatory action, but often in pain or irritation from more transient causes. Even dry heat will often, in these cases, avail much.

PLASTERS.

These are applied to parts either for keeping the spreading edges of a wound in contact and to exclude air, or by stimulative or sedative properties, to impart relief to internal pain and congestion.

The common adhesive plaster should be kept in every family, as it can be applied in many cases by any person of judgment—is itself an excellent application, and there is often urgent call for its use. It both brings together the edges of the wound and aids to protect it from the air. It may be made by adding to a little litharge, melted over a slow fire, about one-sixth of the quantity of powdered resin, but can be purchased more cheaply of the druggists. Directions as to the mode of its application have been given under the head of wounds.

Court plaster can be made by taking a thick solution of gum arabic water, to which about one-fifth the quantity of gum tragacanth has been added, and smearing it evenly over a piece of silk, nailed smoothly down once, and again until sufficient has dried upon it. Then, by being slightly moistened, it may at any time be employed protecting the sore from the air, and irritating soaps or articles of any kind.   .

Stimulating plasters are of various kinds, and often of much benefit.

The mustard plaster, when properly made, is the most efficient and best. To prepare it, take a table-spoonful of ground mustard, more or less, according to the extent of surface upon which you wish to apply it, and add to it water slightly warmed, until you have it the consistency of a common paste; then spread it evenly over a piece of muslin, and apply it to the part. It should not be left on to blister, but long enough to cause considerable smarting and a lively redness. The patient, if old enough, will generally desire it removed soon enough. For children, a little flour may be mixed with the mustard, and a thin gauze of crape or piece of book-muslin, may be placed over the plaster next to the skin, to prevent it adhering. Care should be taken to remove it as soon as the surface becomes quite red, as it will sometimes act in a very few minutes. We often hear it said that the mustard will not draw, and this is attributed to something in connection with the disease, but it is usually owing to poor mustard, too much flour, or the plaster not being pressed evenly upon the skin, so as to come in contact with it. Where no gauze is used, care should be taken, when the plaster is removed, that none of it be left adhering. When an attack is sudden and you desire a speedy action, quite warm water or vinegar is best to mix it, but in a dull, obtuse pain it had better be prepared cool and not so strong, that it may remain the longer. A fly blister does not draw so well upon a surface from which mustard has just been removed.

The common ginger, or ginger and pepper, as we commonly obtain it, mixed with sufficient lard to make it spread evenly, is a very good stimulating or counter-irritant plaster for children in sudden colds, and can be managed and kept on more easily than the mustard.

Horse-radish leaves, dipped in vinegar or warm turpentine, or snuff sprinkled over lard, will be of service where mustard is not at hand, or where it is especially irritating to the skin, as is the case with some persons. Snuff or tobacco in any form should not be used very freely as an application for children. With some physicians, salt pork, sprinkled with black pepper, is a favorite stimulating application.

STRENGTHENING PLASTERS.

Under this name a variety of prepared plasters are sold in the shops purporting to have peculiar tonic and relieving properties, but in general they have no private or special virtues. Where there is a constant pain or weakness, by sticking close to the skin, they produce counter irritation; by causing perspiration or oily moisture of the surface they relax the pores, and by their own stiffness lead the person to favor the particular part to which they are applied, thus providing that rest which is often an excellent medicine for a weak portion of the body. It is mostly, then, their mechanical effect that accomplishes good; but, like poultices, they may be made stimulating or anodyne, or in some other way medicated to suit particular cases. They are most serviceable in chronic affections, and shoemakers' or grafting wax has often been found to be a good poor-mans' plaster; or more elegant ones may be made by combining Burgundy pitch or resin with some oil, or with such medicine sprinkled in with it as will impart either more stimulating or quieting properties.

LINIMENTS.

Liniments form another class of external applications which are often valuable in the treatment of disease. An old prescription for a chill is just before it comes on to run one half mile to a hickory tree, and clasp your hands about it, and run back again. The tree or the hickory had nothing to do with the cure, but the mind thoroughly employed with the race, and the perspiration excited by the action would often ward off the attack. With liniments, it is often not so much the kind of liniment used as the rubbing and friction accompanying their application, and the direction " well rubbed in " is often the best ingredient of the compound. Besides this however, liniments do have some special powers. Their chief object is to stimulate the vessels about a swollen external part, to remove the superabundance of fluid or solid which has partially stagnated there, or so to irritate the external skin as to relieve internal inflammation. An adjacent surface thus irritated takes up more blood, and thus the previous local congestion is partially transferred to a point at which it can do no harm. The most valuable liniments, therefore, are generally those combining an oil and a stimulant. Where there is much pain, some anodyne may be added, or even substituted for the stimulant. We have already, in considering special diseases, referred to two or three excellent liniments. In pain as in colic, or even in cases of swollen extremities almost any of the *essential* oils combined with an equal quantity of melted lard, linseed or sweet oil, and rubbed gently and continuously over the surface will often afford relief. Oil of cinnamon, cloves, rosemary, origanum, peppermint, and the like, may be used in this way.

The mustard liniment is among the best and most ready of this class. It may be made by adding a half-pint of spirits of turpentine and a gill of linseed or sweet oil to a half-ounce of strong ground-mustard, and after shaking it in a bottle now and then, for several days, and then straining off the liquid, you have the preparation at hand for use. It should be rubbed upon the part until it becomes slightly reddened, and may be applied as often as the circumstances of the case may indicate.

A tincture of Spanish flies, prepared with 95 per cent. alcohol, to an ounce of which three ounces of soap-

liniment have been added, is also an excellent stimulating embrocation. The common hartshorn liniment and
opodeldoc, or soap liniments of the shops, may be used
with a similar object. Where it is needed to keep
up a counter irritation for days or weeks, as is often
very desirable where the lungs are affected, the Croton
liniment is generally to be preferred. One teaspoonful
of the pure croton oil, rubbed with five of sweet or
linseed oil, is generally strong enough. One half teaspoonful of this may be rubbed with a piece of kid
glove over the finger, freely upon the chest, morning
and evening until slight pustules begin to appear, and
then it must, for a time, be discontinued. These pustules form often very numerously, and the itching is intense ; but the itching, burning and scratching all aid
in the effect. With children more caution is required
in the application.

The chief of the anodyne liniments are those prepared by the addition of laudanum, chloroform, camphor or belladonna. A tablespoonful of turpentine and
vinegar, and a teaspoonful of salt and laudanum mixed,
combine both the virtues of a stimulant and anodyne.
A tablespoonful of burning fluid, and one teaspoonful
of linseed oil and chloroform furnishes both a cooling
and soothing application. The camphor liniment can
be made by rubbing down a half ounce of camphor in
two ounces good olive or purified linseed oil. From
these varieties a selection can generally be drawn,
suited to any case in which liniments are indicated ;
and while it is convenient for the physician and patients
to have some direction like these to guide the friends

in their preparation and the mode of their use, it is frequently best to have medical advice as to the propriety of their employment.

OINTMENTS.

Among the various external applications employed, ointments occupy a prominent place. These consist of fat or grease, in some form or other, so combined and modified by the addition either of medicinal agents or of other ingredients, as to suit them to the case in hand. Their first general effect, and often a valuable one, is to protect the parts from the air, to soften the skin, or to provide an artificial covering to a raw surface. By the addition of stimulants, anodynes, or the like, special powers, in addition to these, are imparted to them, exciting the sluggish sore to action, lulling pain, or in some other way contributing to cure. In a simple sore, exhibiting no peculiarity different from an ordinary healthy ulcer, the first named indications are the only ones to be met, and for this, a mild, simple ointment is sufficient, and hence one of the most useful and best of the ointments of the doctor's shop is that known as simple ointment. It is made by melting an ounce of white wax and four of lard together, with a moderate heat, and stirring them until cold, or it may be purchased of the druggist nearly as cheap. This is valuable both in itself and also as the best basis for what we may call

medicated ointments. These latter, for all practical
purposes, may be spoken of under four varieties:

Stimulating,
Absorbent or drying,
Astringent,
Anodyne.

Of stimulating ointments, the most common is that
known as the *basilicon*, or royal. It is prepared by
melting together resin, commonly called rosin, 5 ounces,

Lard, ... 8 "
Yellow wax, ... 2 "

and strain through a linen cloth. This furnishes a very
proper dressing for blisters when it is not desired to
heal them rapidly. The two most important medicated
ointments of this class are the compound iodine and the
citrine. The first is especially applicable in cases of
enlarged glands, and the second to foul or unhealthy
ulcers not showing a disposition to heal. Both of them
are best prepared by the druggist and prescribed by the
physician.

Absorbent or drying ointments are among the most
valuable, being indicated whenever there is considerable
discharge from the sore. Their effect in drying up the
disease and restoring healthy action, is alike simple and
decided. Of this class the chalk or litharge ointments
are the best.

The chalk ointment is easily prepared by heating
equal parts of sweet oil and lard over a slow fire, and
stirring in as much finely-powdered prepared chalk as
they will contain. An exact proportion may be given
thus:

Olive oil, . . . four drams.
Lard, . . . half ounce.
Chalk, ... one ounce.

The *Litharge ointment* is made by stirring over a fire,
1 pound of litharge,
1 pint of olive oil,
1 quart of hot water,
until the oil and litharge unite.

Astringent ointments are such as have a tendency to constrict the part to which they are applied. The chief of these are the

Tannin ointment, made by adding one dram of pulverized tannin to an ounce of lead.

III.—*The Sugar of Lead ointment*, made by pulverizing a dram of the acetate or sugar of lead, and then mixing it thoroughly through an ounce of simple ointment.

IV.— *White Lead ointment*, made by stirring an ounce of pulverized white lead into five ounces of simple ointment.

V.—*Zinc ointment.*—Melt three ounces of pure lard, and stir in it one-half ounce of flowers of zinc finely powdered. This is an excellent drying preparation.

Anodyne ointments are prepared by the combination of any of the usual sedatives or narcotics with lard or other fatty substances. The following are among the best of these that can be applied:

I.— Powdered opium, 2 drams.
Powdered camphor, 5 grains,
Simple ointment, 4 drams,
Well mixed and rubbed upon the part.

14

II.—Tincture of belladonna, 1 dram,
"   "   aconite,   "
Chloroform, 1 dram.
Lard or simple ointment (warmed) 1 ounce.
Mix and rub upon the part affected. To this a little prepared chalk may often be added with propriety, as by its use the liquid anodynes are more easily mingled with the fatty material.

Another soothing ointment may be made as follows: Pour upon an ounce of arnica leaves or flowers one-half pint of boiling water, and let it boil away in a covered sauce-pan to one gill. Then strain through a cloth so as to press the infusion thoroughly out of the leaves. Pour in this two ounces of melted lard and ninety-five per cent alcohol each, and stir in sufficient whiting to thicken it. Then add one ounce of melted beeswax, remove it from the fire, and work it thoroughly with a table-knife. While the salve is being prepared, the warm arnica leaves may be placed upon the painful part.

As anodynes are generally better applied in the form of wash or liniment, it will not be necessary to notice them further as ointments.

WASHES.

Washes, or lotions, as they are scientifically called, are external applications often of much benefit to painful, injured or inflamed surfaces. One of their most valuable and frequent designs is to reduce the temperature of a part, and thus check inflammatory action by controlling one of its most prominent symptoms. Accordingly, for this purpose such liquids are to be chosen, and so to be applied as to secure the most rapid evaporation, as the degree of cold is very much dependent upon this. Cold water, frequently applied, is thus of much service. Alcohol evaporates more rapidly, and this is often, in the form of cider-spirit or rum, a valuable application. Sulphuric ether is still more rapid, and a tablespoonful of this in each half gill of water poured in just when it is applied, will cause a greater degree of cold. Vinegar and water, lead-water, ice or ice-water, chloroform, &c., may be used for the same purpose. The design of these kinds of washes being in general directly opposite that of a poultice, the part should be covered only by a light linen cloth, and this frequently removed and cooled. By these means the blood is driven from the part, and the inflammation often either checked or subdued. In mere chronic cases, pouring water in a small stream occasionally upon the part, is often found of service. Here it acts with a tonic as well as cooling power. But besides this important primary value of washes, they may be either anodyne, stimulating, or drying.

Where there is pain with the heat as is often the case, cold hop-tea, laudanum, tincture of belladonna, camphor or chloroform may be freely used with the cold water, and thus the pain lulled, and the patient made more comfortable. Pain then is the guiding indication for this variety of washes.

Stimulating washes are chiefly applied to raw surfaces where there is an absence of action, or where whatever process is going on it is in the wrong direction. To this class belong sores, which, without any great irritation, still show no disposition to heal, or those from which there is a foul discharge. Here, washes are sometimes to be preferred, either to stimulating salves or to any powdered applications. Among the best of this kind are the following: A piece of blue vitriol the size of a hazel-nut, dissolved in a pint of soft water, or twice as much white vitriol may be used in its stead. Twelve grains of nitrate of silver or lunar caustic to a gill of water used soon after preparation, is also often of service. The black-wash, made by shaking a dram of calomel in three gills of lime water, is deservedly a popular appliance. In small pimples, such as often trouble the faces of the young, four grains of corrosive sublimate dissolved in four ounces of cologne water, and used once or twice each day, will often effect a cure.

In washes employed for the purpose of drying up discharges, the liquor is merely the vehicle of the substance possessing the drying power, and the evaporation of the former leaves this deposited upon the sore or surface. Sugar of lead, tannin and chalk, are the prominent ingredients in solutions having this end in view. A half dram of sugar of lead may be dissolved in a half pint of water, or an equal quantity of tannin or chalk, and the wash shaken before using will be ready for application.

Calamine or the powdered oxide of zinc may be used in the same way. With a correct idea as to the designs which are to be accomplished by all these varied outward applications, and looking upon them not as specifics, but as curative by the possession of certain determinate properties, you will be led to rely upon them more than upon unknown compounds, or where they fail or the case requires better judgment than that of the patient, more skillful advice must be obtained.

INFLUENCE OF THE MIND OVER DISEASE.

Both for the patient and physician it is important that the relationship between mind and matter should be recognized and appreciated as an element in disease. The study of disease involves the study of human nature not only in its physical, but its moral phenomena, and one of the most important ideas which the progress of medical science has developed is just this, that medicine is only one of the methods of relieving or curing it, and that other means besides must be combined with it, and other elements brought to bear, both on the corporeal and mental nature, in order to promote and secure recovery. Thus, human will has much to do with pain and suffering. This nervous system is a wondrous mechanism, and its diseases are so linked in, and associated with mental states on the one hand, and with our general framework on the other, that, like telegraphic wires it responds to the minutest change in either. To be nervous was once regarded as a disgrace; but it is time, both for the patient and the doctor, to cease viewing the terms nervous and imaginary as synonymous. Rather, we should view nervous diseases as real derangements of material or function ; and while we medicate for them, we must acknowledge and avail ourselves of the principle, that the will can aid in the accomplishment of restoration.

Abundant illustrations, in common experience, might be adduced to show the power of this mental action over disease. Pain is always more intense when our thoughts are upon it ; from fear, the child will really suffer, although it is purely a mental state ; those who at once give up, as the expression is, when sickness seizes them, are really more likely to die than those whose will battles against the destroyer. The invalid with nothing to do or to occupy the mind, is sure to get worse : no one can endure the brooding over their own ailments, and yet all these are but mere nervous elements in disease. How often is it the case in the experience of those out of health, that if, when feeling very uncomfortable something happens to attract their undivided attention, they are for the time relieved, and hence travelling, amusement, friendship and occupation have cured many a real sufferer, because taking the mind from self, they have removed a barrier to recovery, and allowed nature to triumph unembarrassed. Over every disease in which functional derangement of the nervous system exists, the will, the sensations, have an influence. Upon this principle as a basis, rests all the success that quackery has ever gained. Many a sufferer has been cured by the prescription of the veriest pretender, when the remedy itself was of no more value than pulverized clay; but by faith and confidence mental processes have been controlled, and the due ex ercise of will-power re-established. When a sick man goes to a pretender, a mesmerizer, a clairvoyant, a necromancer, a fortune-teller, a sugar-of-milk man, or a witch, and comes home well, it is vain to attempt to convince him that he has fallen a victim to the duplicity of one who knew nothing about his case, for a real cure is often thus accomplished.

The error of reasoning consists in attributing it to a nonsensical mesmeric, or electrical, or seventh son in-fluence, or to the efficiency of some herb prescribed, and the duplicity of the curer consists in claiming causes which his own experience ought to have taught him are all imaginary. It is true, that in some cases it may be right and proper to conceal from the patient the source of relief, but in many more, the patient himself, by un-derstanding the nature of his trouble, will be able to overcome it. Out of a medical class, for the first time listening to lectures on disease of the heart, as a rule known by a personal experience to most physicians, one half will imagine themselves the subject of the difficulty. Shortness and oppression of breath is as frequently a nervous disorder as it is the signal of serious lung af-fections, and to know what are, and what are not nervous symptoms, enables many to overcome the former or at least to give themselves no uneasiness about them. It should be the effort of the invalid and of his physician, so to occupy the mind, as not to tax its energies over-much, and yet prevent that idleness on which melan-choly loves to live. Just, as naturally timid men, like some soldiers and surgeons, have learned coolness by experience, so many a nervous man by determination, decision, will, and the proper use of sanitary dietetic and medical regulations, will control the machinery of nervous action and prevent it from unduly controlling him. The cerebellum and sensorium are two special divisions of the nervous system or brain, the one having the office to give direction to the powers that move the muscles of voluntary motion, and the other recording all sensations experienced, and a large portion of the common disorders of the nervous system consist in a loss of due connection between the cerebellum, or will and motion, and an undue exaltation and acuteness of sensation, and in the treatment of such cases both by the physician and patient, the problem to be solved is, how to place the sufferer under such circumstances as shall be most favorable to physical and mental health, and as shall restore to him that equilibrium which while nourishment and tone is given to the physical frame will enable him to reinstate the will in the legitimate exercise of its functions.

None are more deserving of sympathy than the nervous, and yet none with aid from hygienic and medical measures can do so much by personal determination to rid themselves from their real and inexpressible suffering. The common expression, " shaking off discouragements," has philosophy in it, and ill-health, from which misanthropy often arises, where dependent merely upon functional nervous derangements, may with a little foreign or outside aid often be overcome.

Let the invalid and the doctor appreciate this princi-

ple and both to their intense satisfaction will find by
experience its practical truthfulness.

THE PRESERVATION OF HEALTH.

This, theoretically, is an object worthy of the greatest
personal effort and sacrifice, but practically there is no
temporal blessing so utterly neglected.   It has been too
truly remarked, that "an enormous portion of mankind
are so habitually invalids that they have no notion of
any other state of existence."

Self-denial is required, and this of itself is sufficient to
render the task, at first, difficult and irksome.   It will surprise any one, who for years will make the subject one of
accurate observation, to find how much of disease is the
direct result of personal imprudence.

Original sin we have, and inherited diseases afflict
many; but a careful analysis fully shows that those
otherwise dormant are aggravated by our own mis-
guided abuse, and that a much larger share than is generally supposed, arise from personal disobedience of
nature's laws, or from the wrong training of our immediate parentage.

It should be the business of each one for himself, to become informed as to the plain and sensible laws of
health.   The rules of exercise, diet, ventilation and
physical culture in general, are as determinate as need
be, and the overstepping of them is oftener the result of

14*

ignorance, morbid appetites or uncontrolled desires, than of any uncertainty in the laws themselves.

The first great difficulty is, that the physical home-training of society is loose and incorrect. .

At the outset, of all kinds of family government this is the most important. Animal appetites and instincts declare themselves before the intellectual and moral; and to conform these to the control of human faculties, is the first opportunity we have to teach the infant lessons which impart to it a habit, a nobility of power which in advance, aids in the subjection of the mental as well as the physical. The first effort then of a parent should be, to teach the child ere it is aware, physical habits which will mature into a general self-control when years of discretion arrive.

The tender infant nursed every two or three hours and then not so much as to overload the stomach, is no more troublesome, after a little, than the one which is fed whenever it cries, and cries whenever it is fed, while at the same time it is forming its first good habit.

The small child who has plenty of plain, good food allowed it, but this at regular intervals, and is taught to eat slowly, to deny itself articles positively injurious, to use without stint and yet in moderation those things of which it is fond, is learning a noble page in the volume of its life. The parent who feels that appetites are themselves exposed to excess and do not need pampering, and therefore charms his children with other gifts than those creating vitiated tastes or encouraging gluttony, is laying foundations for a symmetrical superstructure of health and character.

Oh! how much these early physical habits have to do with the future welfare of our offspring, in mind, soul, body and estate.

We have plastic material with which to deal, but it can only be moulded aright in the hands of those who think and study upon the objects and modes of discipline. Children are the quickest of learners, the most perfect of imitators, and much of what is called heredi tary in disease, in character and in action, is acquired after birth from those from whom the child makes copy. Wonderful is the power we possess to determine what they shall be, even on the score of health. Referring to physical management and discipline, Dr. Beale, an able writer on this subject, makes the following just remarks: "A long-continued course of injudicious feeding, want of air and exercise, indulgence of various kinds, and neglect of bodily and mental activity, will effect such a change, that an infant born of healthy parents, with all its organs well fashioned, may become a miserable, rickety, scrofulous child. The reverse of this also occurs. A delicate infant, born of weakly parents, may, by a very judicious and long-continued system, become a healthy and happy child.

The most miserable example of scrofula may, by well-directed means, by attention to all the laws of health, in the course of time, become the picture of good health. The period required for these changes is often very considerable; nature effects them by a slow and almost imperceptible process. Those who would recover health from a state of chronic disease, must be content to persevere in a right system, for many months before the constitution, in all its parts, can be made to work well.

There is no royal road to health; and if a constitution be weak by nature, or has become so by untoward circumstances, the production or renovation of health can only be effected by a very determined perseverance.

The public, in general, entertain such false notions of the powers of medicine, and their faith is so akin to superstition, that the majority expect and believe that there is a specific remedy for every disease, which, if unknown to one, may be known to another practitioner, and they run from one physician or surgeon to another, never trying any means proposed, long enough to know whether it will do good; thus they become confirmed invalids and the dupes of every quack within their reach. Whereas, if in the early stage of disease, they would consult a medical man who deserves their confidence by his knowledge and integrity; if they would be content to pursue health as nature dictates, by laws easy of comprehension and undeviating in their results, they would gradually recover that vigor which is not to be obtained in any other way." Judicious management often works wonders with disease, and perfect self-control of the physical once attained is a possession; itself a greatness.

There are two extremes frequently followed in the early care of the health of children. One class leave everything, as they say, to nature. The child runs at large; eats what it pleases and when it pleases; is dressed to suit convenience or fashion, in utter disregard of comfort; sleeps or plays at pleasure; is in or out of doors according to its choice; in a word, instead of being "brought up," grows up like a neglected hedge. For those of naturally strong constitutions, tough, robust, hardy, this course often seems to answer very well. They weather these early life-storms, and parents often point to them as the proofs of the wisdom of their system of government. Yet, even on them, imprudent exposures, unrestrained appetites, uncontrolled passions, leave their deep furrows; and if they survive the earlier and ruder blasts of life, a worn-out stomach is the first token of approaching age.

For the "feeble folk," this plan is a decided refinement upon the old Spartan one, for they killed the delicate children outright, boldly avowing them not to be worth raising. This system destroys them more slowly and genteelly, but no less effectually. The statistics of those who have carefully collected and observed the aggregate of facts furnished by this method, show that the health of the living is not a fair basis for computation and conclusion; that a much larger proportion than is supposed fall by the wayside, and that tribes, such as the gypsies, Indians, and others who have followed out the system, have become decimated or extinct.

Another class fall upon another and no less objectionable extreme. Their children are disciplined with all the strictness of militia in service. They eat only the things laid down in the books as digestible; drink even at breakfast, in the coldest of weather, nothing but the coldest of water; sleep between feather beds; are restrained from romp and play, because it is rude; sit on high chairs like statuary, since this is what is called "behaving nice;" are fed without any reference to the kind of food or the amount of exercise, but just so much; are kept as much in the house as possible, and are early taught to be afraid of " out of doors," while to occupy them within, and that they may be early accounted *smart*, their minds are crammed and stuffed to precocious fullness.

It needs nothing but common sense, reflection, and a fair degree of knowledge as to the operations of human existence and the laws of health, to enable every one to attain a reasonable medium betwixt these extremes. Substantial food regularly taken, clothing sufficient to secure comfortable warmth, dry feet, frequent bathing, rooms well ventilated, plenty of exercise in the open air, and early attention to the first symptoms of disease, are all so in accordance with the dictates of good judgment and prudence, that their value ought to be appreciated by all.

The child taught to eat slowly, and provided with good and plain food, will seldom eat too much ; and if well clad and protected from early dews, rains or intense heat, will seldom exercise too much; and the great point of physical training is, not to supersede natural inclinations, but to direct, modify and restrain them by more mature judgment.

The foundations of health are laid sooner even than those of character, and have a relationship to it; and for the first seven years, intellectual and moral culture are secondary to physical.   Next to family management the school-room and school exercises have much to do with health, and that is an epoch in life when a child is brought under an influence second only to that of home. A well-ventilated room, pleasant seats, an easy posture, and a proper admixture of play with study, is just as important as that the teacher should be an adept in what has been known as the three (r's) of "reading, 'riting and 'rithmetic." Here, too, should be early taught that knowledge of this human self, and of the laws of health which will impart correct views for future guidance. How many a one to-day in ill or good health, is indebted for it to the ill or good physical training of teachers and parents.

Animal appetites and passions, and acquired propensities fastened by the habits of early life, are often stronger than the maturity of reason, and hence we can scarcely over-estimate the importance of diligent restraint.

But taking those of us, in adult life as we are, how shall we fulfill the laws most conducive to the preservation or restoration of health. Study and obey them is the easy answer, but the actual practice, the sternest of mental acquirements. If life's early tuition has been correct, the task is much more easy, but if not, the attainment is worthy of the effort. The embarrassment is not in their comprehension, for they are easily understood, but energy and self-control are the masterly prerequisites. You must, by examination and experience, learn to appreciate their justness, and to feel that comfort, health and the endurance of life itself depends upon a proper regulation. If indisposed to exercise, place yourself in a position or avocation which will make it a necessity, if overwhelmed with business-cares, remember that property is worth less than health, and act as if you believed it, if a " man given to appetites" have only proper food brought before you, until you are able to endure the sight of viands, pastry and confectionary, and still enjoy the better luxuries spread upon your tables, and thus in every way avoid temptation, until you gain that mastery of self which, from being self-denial becomes a pleasure, laden with manifold rewards. Many a man who has dared the cannon's mouth, nor feared to speak bold eloquence to kings, has quailed, subdued by tables filled with luxuries, or fallen victims to the lulling sway of wealthy indolence.

Many a conscience susceptible to other moral sentiments and obligations, seems wholly blunted as to daily physical sins, and the moral obliquity of neglecting one's health is a wickedness for which, in common, priest and patron pay enormous penalty. More than we are apt to feel, a kind Providence has placed our lives and healths in our own keeping, and neither in the paths of religion, of literature or of art, was pain meant to be the rule. Let each one of us see to it, that we study the laws of " this fearfully and wonderfully made" organism, and both as a pleasure, a profit and a duty, do our part in preserving the health and integrity of this "human form divine."

Beside the care of our own health, and that of our children or those committed to us, the public health of society at large, has much to do with our own personal protection from disease, and commends itself to our attention by the still higher claims of an earnest, far-reaching, race-loving philanthropy. Even from the stand-point of political economy, sickness is a sore national evil. Every sick man is a direct burden upon the productive capital of a country, and commerce, manufactures and industry of all kinds, enter a deficit on the day-book of loss.

In cities, benevolence is too exclusively directed to providing for the suffering rather than abating or mitigating the cause, and much of the capital spent for hospitals, houses of refuge, asylums, and penitentiaries, would be more wisely appropriated if expended in preventing the necessities for their existence.

A quarantine, while it keeps out disease, must see to it that by locating and congregating the poison, it does not render it by its mass and intensity, more virulent, and a still higher and almost superseding step of prevention, is to address chief attention to the previous sanitary condition of ships, passengers and crew.

The condition of tenement houses calls as imperiously for the surveillance of a special police, as does the hospital for a medical staff; a city can afford public laundries almost or quite free, and then compel cleanliness, more profitably than it can support dispensaries or pay in other ways its yearly thousands for the results of filth and carelessness.

Public baths, every few hundred rods in the poorer districts, though it might seem like going back to Roman civilization, would do more for the health of a metropolis than human calculations can compute, and monies expended in the proper removal of human excrement, and vegetable decay, in chloride of lime, plaster, or other disinfectants, and in applying various well settled hygiene regulations would either pay their own adequate percentage or inflict a tax much more bearable than that accruing from manifold less important sources.

In the country such regulations must have chief reference to miasmatic localities, and in diffusing a knowledge of the laws of health; and that was a wise measure of political economy in enlightened England which appropriated large sums toward aiding the farmers in the draining of their own private lands. A nation which directs munificent attention to the health of its subjects gives wise evidence of advancing civilization and refinement.

We have seen even in our brief notice that the preservation of health is a grand problem, involving important questions and principles in personal and home influence, in education, in public interests in life, in happiness, in all that relates to the social, moral, intellectual, and national prosperity of the world, and yet only needing a careful consideration in order to secure for it, in many particulars, a solution far more satisfactory than present experience illustrates.

In this as well as in other subjects treated in this book, our aim has been to direct attention to principles of physical health, so as to introduce them as topics of thought rather than follow them out in all the minutiae of detail; and to notice practice only so far as may be comprehended by all or may be rendered necessary by the exigences of sudden attack.

In the first effort we have relied upon what we believe to be the first value of a book, to lead men to think for themselves on the theme treated, so that the author shall be the momentum and the reader the developing force.

In the second, we have felt that the main dependence of the sick must be the attending physician, with sound learning, good judgment, and faithful experience, watching the case, but that a manual like this would guide in avoiding or detecting disease, in doing rightly the something which nature prompts all to do until a medical attendant is obtained, and assist in enabling friends to fulfill directions and in providing such management as is often essential to successful treatment and complete recovery. It is the highest aim of our profession, to prevent, relieve or cure disease, to support the weak, soothe the suffering, ease the dying, and extend the hand of sympathy and help to humanity everywhere, and we trust these brief pages will assist both physician and patient in preventing, when apart, errors in management which would prove injurious, and that they will lead both more highly to value that relationship around which group hope, friendship, happiness, and all of earth that is dear and precious, since health and human life itself entrusted to our care, and encircled by solemn anxieties, are watching and awaiting the result.

According to the idea of the old Greek philosopher, "to know thyself" is the perfection of all human knowledge; and surely, individual man in himself considered, is a theme of study full of interest, and one the wonder of which investigation cannot exhaust. Viewed simply in his material and physical arrangement, he affords an illustration of mechanism, design and adaptation which imparts fresh ardor to every inquiry, and is well worthy of our attentive regard. It may not be uninteresting to glance at a few of these organs on which our life and comfort depend, and to notice the methods by which our maintenance and happiness are secured.

By means of our senses we are brought in contact with the external world, and these are thus made the mediums both for receiving and imparting knowledge. Curious and wonderful is the perfection with which every thing in respect to them is arranged. First, as a basis for their action, we have the nervous system, the centre of thought and voluntary motion, firmly encased in the arch of bones which forms the head, and in that column of movable strength and careful protection which is afforded by the spine. From these sources spread out those nerves which by an inherent power imparted to them by the unseen Maker, are prepared to appropriate the materials of knowledge transferred to them through the medium of the senses.

How perfect, admirable and wonderful the physical arrangements by which these results are secured. Could you place a small portion of the skin which covers your body under a small microscope, directly beneath the outside surface and its coloring matter, you would find small eminences thickly studded with the points or loops of minute nerves, so close together that you cannot thrust in the point of a cambric needle without touching them. The effect is instantly communicated as by a telegraphic despatch to the brain, and the mind stirred to responsive action. In some situations the arrangement can be seen, as by tracing the minute line-marks upon the palm of the hand; and when we remember that every touch at any part of our body is recognized, recorded and responded to with a quickness which, for the want of a better term, we call quick as thought, we cannot but be delighted with the proofs of design thus afforded.

By a similar arrangement upon the surface of the tongue, and by the commingling of nerves both of taste and touch, we have secured the ability to distinguish all varieties of flavor, and thus to choose the pleasant and reject the disagreeable. Then, again, in the nose we have an organ so constructed that the branches of the olfactory, or nerve of smell, are spread over an extended surface, and we thus enabled to detect odors beneficial or prejudicial to human life. But it is especially in the organs of sight and hearing, that the perfection of mechanical arrangement is exhibited, and we may therefore more minutely examine their sublime contrivances.

If light is the wonder of the outer, inanimate, material world, no less is the eye the centre of the grandeur of the inner man, world of the human soul and life.

The mechanism by which it is adapted to the waves of color which vibrate about us, and by which ability is imparted to it to receive and distinguish impressions, and its transcendent powers of easy motion, the more closely they are examined have a tendency to stir the intellect into holy and admiring enthusiasm. As we behold it, it is a small sphere set in a nicely rounded cavity, with another sphere inserted into it, and this in its centre penetrated by a point of ebony, a mirror ever changing with the different rays, and painting with more than artists' skill each hue presented to its gaze.

This sclerotic, or hard, whitish coat which constitutes four-fifths of the external coating of the eye-ball, is plentifully supplied with nerves and arteries. In it the cornea or middle portion of the outer eye is set like the watch-glass in a groove fitted to receive it, and it is indeed the transparent centre through which alone the light can enter. Next to this is what is known as the aqueous humor, a little fluid encased between the cornea in front and the iris and pupil behind. It is this iris or rainbow which imparts the different color to the eyes of different individuals. It is penetrated near its centre by the circular opening called the pupil, and has the muscular power either of dilating or contracting the size of this aperture, and thus of opening or shutting the window of the mind according as the power of light may direct. Directly behind the pupil is situated the crystalline lens, embedded in a transparent fluid known as the vitreous or glass-like humor, but separated from it by its own membrane. This lens consists of transparent layers of membrane, each one from without inward harder than its adjacent one, and in the centre a few drops of fluid.

The third portion of the eye is the retina, the indispensable of sight, the plate on which the image of every object seen is engraved, the network expansion of the optic nerve in which the power of seeing is located.

By this arrangement or series of arrangements, we have in the human eye the most perfect and perfectly arranged of all optical instruments. The rays of this subtle element, light coming from any object upon the cornea, by its shape and density are made slightly to approach each other. These pass through the aqueous humor until they come to the pupil, and its muscles fit its size to the degree of light. The rays pass on to the vitreous humor and the crystalline lens, and are still more bent or refracted until they meet in a focal point at the retina, and an image of the object seen is thus accurately pencilled. This retina is spread out over a black surface, so that the rays are prevented from being reflected again, which would 'cause a confusion of sight. The perfect optical instruments of the present day are but imitations of this most perfect of all, and many of the inventions in optics which adorn and illustrate the present age, have been directly derived from a study of the eye. The crystalline lens, the convex cornea, layers of different thickness and density, black surfaces to absorb the rays, and fluids of different densities, properly to bring the light from the object to a focus, are the most closely imitated in the most accurate instruments of the astronomer, the microscopist and the daguerrean. By arrangements no less wonderful, its exquisite powers of expression and motion as well as of sight are secured. By a mode of attachment exhibited in the accompanying figure, the eye moves responsive to muscle fitted to draw it upward or downward, right or left. Two of them acting together draw it to intermediate positions,

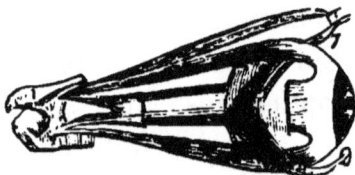

all united in power it is actually drawn further back within its orbit. The two oblique muscles, one of them acting through a pulley, roll the eye in any direction, while these together have a power of drawing it slightly forward, and thus a facility and accuracy of motion are secured well adapted to impress us with adoring and delighted wonderment.

THE SENSE OF HEARING.

The arrangement by which the power of perceiving sounds is secured, though not so impressive in its outer conformation as the eye, will be found, even on a superficial investigation, no less accurate and skillful. Its external portion, though not the subject of the poet's or lover's reverie, still has an important office to perform, and is even more an ornament than we are apt to imagine in these days when ear-cropping for crime is out of date. By its varied shape it is well fitted to catch the waves of sound as they go trembling through the air and convey them to the more delicate organs within. A tube of less than an inch and a half in length, wider in its middle portion than at either end, leads to a delicate membrane stretched over the termination of this short tube, not inaptly compared to the head of a drum. Next to this is a small room hollowed out of a solid bone like a cone in the midst of a firm rock. Within are contained three small boxes, known as the hammer, the anvil and the stirrup. The hammer is fastened to the inner central surface of the membrane before noticed, and by its other end to the anvil, and this again to the stirrup, so that the three form a series of bones attached to each other, and yet movable by means of muscles and ligaments attached to them. The further end of the stirrup is attached to a muscle and in contact with the membrane which connects with a cavity still deeper in the skull. The following figure exhibits this wonderful arrangement.

But it is necessary that air be admitted to this cavity, and from one of its sides we accordingly have a small tube known as the Eustachean, extending to the throat or pharynx, and opening thereinto. Behind the tympanum or middle ear and more deeply imbedded in the rock-like bone, we find still another department, if

15

possible more delicately and wonderfully chiseled than
the former. So intricate and devious are its windings,
that the old anatomists called it the labyrinth. The en-
trance-hall thereto, directly behind the tympanum, they
named the vestibule. From it jut out three long tubes
which circle into it again, and are hence called the semicircular canals. Then, beside all this, there is a curve
of bone which turns two or three times round a central
axis, and is pierced with numerous minute openings for
the passage of fibres of the nerves. This, from its shape,
is called the cochlea or snail-shell. Corresponding to
all this, there is a softer membranous labyrinth filled
with a sparkling, limpid fluid, and the branches of the
nerve of hearing are freely distributed throughout this
mystic cave. The following diagram shows the arrangement of this internal portion of the hearing organ. It
is said that when the anatomists, after admiring the middle ear, first sawed into this portion and found this unknown deeper cell, so overcome were they with wonder
that they broke forth in impromptu language of praise
to the glorious Maker of us all.

Now hearing, so far as we know, is accomplished simply thus: The waves of air or sound striking upon the
external ear, are first conveyed to the tympanum membrane or drum. Under the impressions thus made, a
motion occurs in the row of small bones we have already
noticed, by which the sound is conveyed still further on.
Into this middle room air is constantly admitted through
the tube from the mouth. Thus the drum membrane is
pressed upon by air from either side, and beside, an opening is afforded through which air can escape when heavy

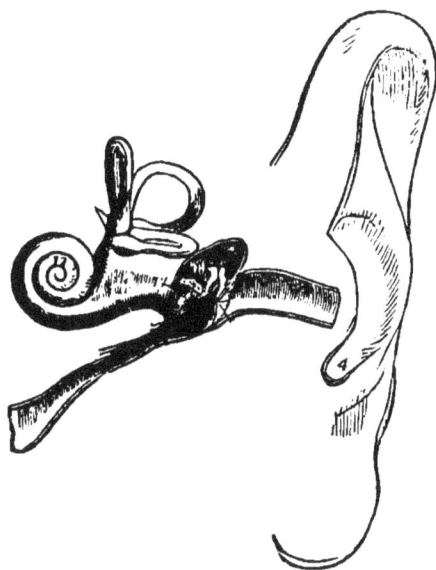

sounds are produced, and hence it is natural quickly to open the mouth when a loud and near report is made. By the action of the series of small bones, the ear membrane, like the pupil of the eye, is expanded or contracted under its over-stimulus of sound. The sound conveyed by these bony messengers through the air-chamber, comes at last and knocks for admission into the labyrinth. Here, upon spiral windings, and semicircular canals, and convolutions of membrane and bone, are spread out the nerves to receive the sensation of hearing; and that this may be more accurate, a spring of water pure and clear is there, rising and falling to the action of the outer waves of sound. In this liquid are suspended three wondrous crystals of carbonate and phosphate of lime, with which the nerves for sound connect.

The depth of the mystery we cannot fathom, but enough we can see and know, to perceive how accurate is that mechanism which enables us to distinguish and arrange sounds, which feeds us with the delicacies of musical notes, and detects a less than hair-breadth variation from correct execution of the noblest productions of the noblest composers.

Thus the machinery of the senses, intricate yet grand, is thoroughly elaborated, and so nicely do taste, and touch, and smell, and sight, and sound perform their appointed tasks, that amid all their abuses, in the vast majority of cases each organ fulfils its appointed part. All of them can be improved by cultivation, or injured by abuse; and in these days of vitiated tastes, over-brilliant lights, and strange table-sounds, we must take care that in the use and management of all, the great sixth organ, common sense, presides.

ORGANS OF CIRCULATION.

Of these, the two chief centres are the heart and lungs, while the arteries and veins are the tubes through which their action is transmitted to other parts. The heart is nothing more or less than a receiving and distributing reservoir of the blood of the system, having both the power of receiving and of thrusting forth the fluid coming into it. Its usual weight is about eight ounces, and it is divided into four apartments. These are known as the right auricle and ventricle, and the left auricle and ventricle. Now the course of circulation is simply this: The blood from the veins having become dark and more impure than that in the arteries, by taking up spent and noxious material as it has traveled through the system, pours itself by the large veins into the right auricular apartment. By a slight propelling power of the auricle, this flows into the next cavity, the right ventricle. From this, by the beat of the heart, it is pumped up through a large tube into the lungs, where, penetrating each minute vessel, it is brought in contact with the air we breathe, losing its carbonic acid and receiving pure oxygen in its place. With color thus changed, and self thus purified, it flows on to another large tube, and as arterial blood enters the left auricle. From this, through a neat valve, it passes to the left ventricle, and from it, through a large vessel, is pumped out into the smaller arteries throughout the whole body. This, circulating through the most minute tubes, is gathered into the veins, which in its impure state convey it to the right auricle, again to undergo the purifying process.

THE LUNGS.

These are two organs situated one on each side of the chest, whose office is to receive pure air, and thus purify the blood, and to breathe out the foul gases of the system. First we have a large tube, the windpipe, leading from the mouth. This, extending just within the cavity of the chest, divides into two pipes, one for each lung, known as the right and left bronchial tubes, the right, the larger, passing off nearly at right angles, and the left more obliquely. These each soon divide into two more, and thus keep dividing and sub-dividing so that they number thousands; and an inverted tree with its

multitudes of branches conveys some idea of their
divisions. These finally become excessively small, and
terminate in air-cells. Amid all these, blood-vessels
equally minute are distributed, and the fluid in them so
affected by this flowing stream of air, as to give up its
impurity and take something valuable in its place. So
the lungs may be said to be a network of air and blood-
vessels, other tissue serving merely to connect these.
This figure gives a correct idea of the size and relative
positions of the heart and lungs, H representing the heart,
and R L and L L the right and left lungs respectively.

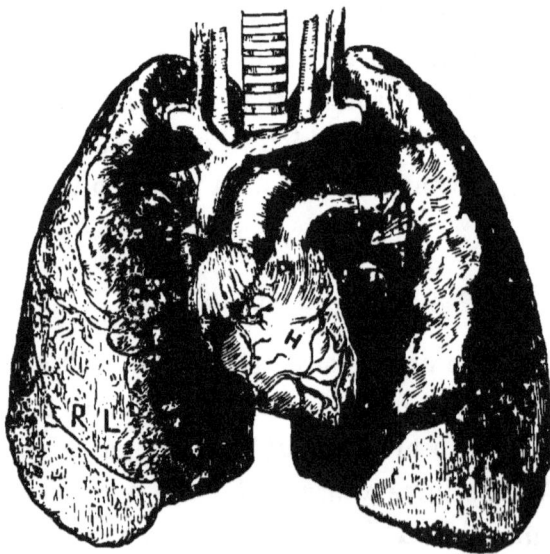

Beneath the lungs and heart is a broad sheet of membrane dividing the cavity of the chest from that of the
abdomen. In the latter are to be found the stomach
and intestines, the liver, pancreas and spleen, and the
organs for separating the urine.

The relative position of some of the more important organs, both of the chest and abdominal cavity, is shown in the annexed diagram.

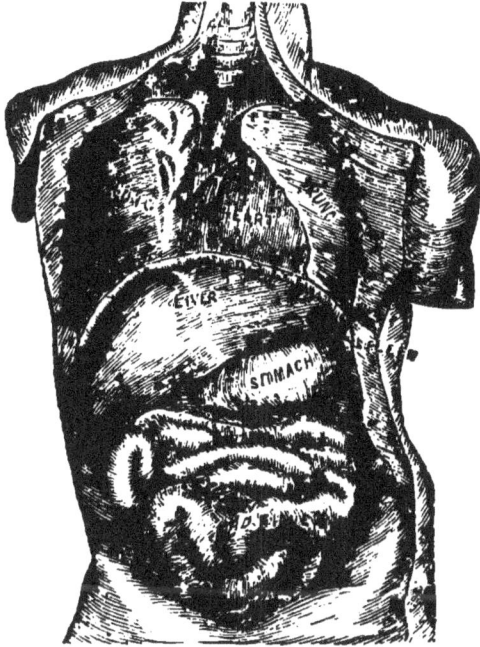

The stomach is the receptacle of our food, and the first and chief organ of digestion. The œsophagus or food-pipe enters it a little to the left side of the middle line of the body, and the general shape of the organ is shown in the figure.   (See page 344.)

From this the small intestines, about twenty-five feet in length, extend to near the right hip-bone, where they expand into the large intestine, which, after a circuitous course of about five feet, ends in the rectum.

The liver, the largest organ in the body, usually weighing about four pounds, is situated, as shown in the

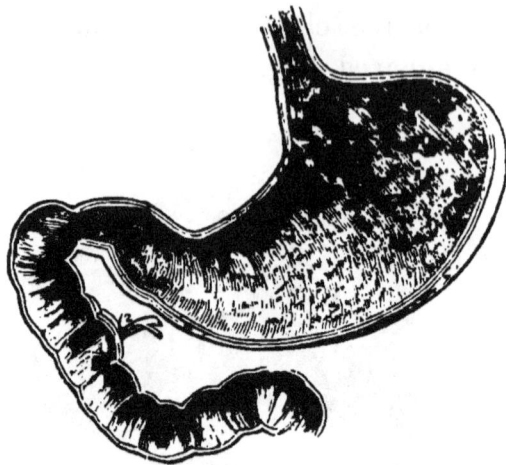

large figure (page 343) chiefly to the right of the middle line, but extends also across to the left side, measuring in its longest diameter about one foot. It aids in separating impurities from the blood, and secretes bile. This is discharged by a small tube into the intestine, just below the lower opening of the stomach, and performs an important office in connection with digestion. The gall-bladder is a receptacle of bile, and is connected with the intestine by the same tube.

The pancreas or sweet-bread, as it is sometimes called in animals, is an organ lying behind the stomach, usually about four ounces in weight. It performs some office in connection with digestion not yet fully understood.

The spleen is situated in the left side, near the greater curve of the stomach, is oblong in shape, and weighs, when not distended, about eight ounces. It seems to be a kind of safety valve of the system, formed like a

sponge, so as to receive a large amount of blood in case
of obstruction elsewhere, or when the stomach is distended with food.   When our aliment is received into
the stomach, by the action of the juices and secretions
thereof it is all reduced to a semi-fluid state.   A part
of the nutrient materials are abstracted, while the
starchy compounds and oil pass on in some way to be
affected by the alkaline bile and pancreatic juice.   All
that is capable of affording nourishment is taken up by
small glands or tubes in the intestines, while the rest
passes on to be rejected.   The materials taken up by
these vessels of the stomach and bowels finally all flow
together into a tube about the size of a goose quill,
which empties into one of the veins.   Thus a fresh
supply is constantly granted to the system.

The kidneys, situated one on each side, in what is generally known as the small of the back, are two bodies
of oval shape, about five inches long, two broad, and
weighing about four ounces.   By some peculiar power
they separate the urine from the blood as a noxious
material, and by means of small tubes it trickles into
the bladder, an organ prepared to retain it for a time.
We thus see that in the chest and abdomen we have
provided, a propelling power for blood, an organ by
which air to purify it may be conveyed to it, a digestive
apparatus furnishing constant supplies, glands secreting
necessary acids, other separating what is unnecessary or
injurious, and still other providing depositories for spent
material, so that even in an imperfect survey we may in
some measure comprehend the wondrous mechanism,
chemistry and philosophy of man.   Bring together all

15*

the science of ages, the noblest theorems, the grandest problems, the most sublime embodiments of artistic skill, and yet in-one's little self may be found exhibitions more wondrous, adaptations more useful, appliances more wise, and results more inspiring than any afforded elsewhere.

.

SHORT INFORMAL LETTERS TO VARIOUS CLASSES.

Letter I.

To Parents and Teachers.

The responsibility of bringing up and training children, is even greater in practice than in theory. There are but few of us who do not at times feel the greatness of the charge in its bearings on intellectual and moral culture, but many, I fear, do not appreciate the importance of proper physical regimen and development.

Health is a blessing so appreciable, that mistakes as to the management of children ought to be most sedulously guarded against—for who would not shrink from entailing in any way upon his offspring, or those committed to his care, evils which scatter pain and suffering along life's journeying, and make those we love the recipients of hard penalties at our hands? Each child, even in its physical interests, is a book to be studied, a theme to be investigated, a subject to be considered, a problem to be analyzed. As there are modes of discipline adapted to various mental and moral proclivities, so there are differences, too, in the training essential to various physical conformations, and to these reference must be had, if we would confer ruddy health and firm endurance upon those whom we are preparing for the stern conflicts of life. This, difficult as it may be, is easier to be understood than the deep things of mind and soul; and as to no part of us is nature so outspoken, as in respect to our physical wants, in no department will common reason and common sense lead to such sensible and practical results; and yet, in the application of these laws we are all of us more or less sensible of partial failure. Says Dr. Combe, "Those whose opportunities of observation have been extensive, will agree with me in opinion that at least one-half of the death's occurring during the first two years of existence, are ascribable to mismanagement and to errors of diet." Now this includes a period when, of all others, it would be thought our indications are the plainest. Milk, and nothing but milk, has been provided for the first division of this time, and yet very many of the parents are dissatisfied with this arrangement. One class wean the little ones outright, on the ground of convenience. The Zulus in South Africa, when twins are born, kill one of them on the same moral argument. We are shocked at their barbarity, but have abundant apologies for those who expose their little ones to the tender mercies of spoon-victuals, and to a prolonged, but in many cases no less certain doom.

Nature, too, as a rule, is plain in her indications as to when change to solid food is desirable. The first teeth are not teeth for chewing meat, but merely for dividing food requiring no mastication; while in due time the grinders appear, ready to do their part in the offices of life. Then as to quantity, among those who talk most of the danger of starving the dear creatures, we have never seen a case approximating to such reduction. We are no advocates for staid rules of half-gills and ounces, but we do believe there is a natural appetite, which even in the child limits itself at first, and which will continue "sensible, as instinct always is," unless perverted by mismanagement. The new-born infant will cease nursing when it has enough; but if allowed to lie at the breast, if coaxed to imbibe more at every cry, it will soon learn its first unfortunate lesson. Then, again, the properly managed child, when at first fed, will limit its desire by its feeling of full satisfaction, and indicate its enough, unless coaxed to more; and the older child, taught to eat slowly and chew well, will seldom eat more than it needs. Our safest rule in the start, in respect to children, is to see to it that early appetites are not pampered; for this we presume for them a safe guide in the future. I know many families who allow the child at the table only at breakfast and dinner, giving it as much of a plain tea as it may call for, but not sitting down to a table of luxuries, the stuff of dreams and sleep full of unrefreshing agonies. If some children could tell us what they dream, after such doses, methinks we would not wonder that with brain and bowel ailments half the children die.

Then, again, as to the subjects of heat and ventilation, all are aware that all we need is to be kept comfortably warm; and yet with this plain fact before us, and cheap thermometers to test the exact amount of heat, how many suffer from uneven temperature! The pleasure, necessity and benefit of pure, fresh air, is readily agreed by all; but yet, children are huddled together in close apartments, and the plain laws of health subverted oftener on grounds of convenience, than of necessity. The benefit of lively out-door activity we all admit, when we gaze on a ruddy clan of farmer-boys—but how much, too, is this neglected. Even in the confines of a city lot, there is room for ball and battledore, and many other little stirring games, and catching cold is oftener done in shut-up nurseries than in active play. Very, very seldom does the romping child become sick from cold, if coming in when the play is over.

The teacher, in his noble sphere, has ample opportunity to inquire into the physical necessities of his pupils. Is the room well ventilated? are the seats comfortable? is the position erect? are the intervals for play sufficient in length and frequency to promote rest and refreshment? Questions such as these need but be suggested to the intelligent, in order that they may appreciate their importance, and the requisite culture is readily maintained. If, as is your aim, you would secure the full training of the mind, the shell must be sound, or it will harm the kernel. The sap of intellect circulates through bark, and branch, and leaf, and stem; and a good mind in an invalid body, is too often a giant struggling in vain beneath the fallen pillars of a Parthenon. May you, my dear readers, be you parent, or teacher, or in any way a director of youth, not forget to inquire what it is even physically to train a child aright.

Letter II.

To Students and Professional Men.

No class of our race are at the present day doing more injustice to themselves and to community, than those engaged in literary or mental pursuits. It would seem that their very intelligence and education should be the safeguard against the dire neglect of the conditions of health, and yet sad facts show a strange oversight even of physical axioms. The history of literature, of science, and of mental acquirements, is too much a history of invalids, and there is some propriety in defining scholars as "complaining men who eat much and exercise little." The love of books and of learning is indeed a noble passion, but the enticement of the charm must not be allowed to hurry the body into premature decay. The want of regularity and of proper lively exercise, in which subjects of study are dismissed, and even the mind given up to the sway of its present occupation, is the fruitful source of educated ill health. None is more to be pitied than the intellectual sufferer, the mind rendered sensitive and acute by training, and the body a broken constitution, like a shattered musical instrument resounding plaintive discord, sad accompaniment, or rather, sad impediment to the noble executions of intellect, the more sad because too often the direct result of injudicious confinement. Study is continued until an exhausted mind disposes the body to a slothful inactivity, and exertion is avoided, not so much for want of time, as for want of inclination. There is no greater error of logic, than the oft-repeated excuse of the student, that he has not time for active manual employment. Statistics have shown that he who observes the Sabbath will accomplish more than they who labor every day alike; and equally true is it, that the mind is capable of more, when the body which contains it is kept in working order. The buoyant step, the active walk, the jolting ride, the hearty sport, is not time lost from the studio, but a guarantee that the whole frame-work will return to the task more able to compete with the intricacies of ancient or modern lore. Self-sacrifices on the altar of learning, are no less venial than other; and neither our Maker nor the nations ask or need of us devotion to study so intense as to destroy health. We have often been surprised, however good men shelter their sin under the most flimsy excuses. Recreation is as much a duty as it is a privilege—makes achievments mentally as well as physically—and the biography of greatness proves that most of those whose names are plainly legible upon the scroll of fame, have written the productions which have made their names immortal, by improving the hours when they found their minds exuberant, active or analytic, rather than by goading them on with lengthened measurements of time. As to eating and drinking, the student and the learned incline to two quite opposite extremes. The one class, with appetite inflamed by unreasonable indulgence, or rendered morbid by nervous sensations resulting from an overtaxed brain, have plunged into the excesses of gluttony and intemperance, while others more safely but yet erroneously have landed on the shoals of ab-

stenuiousness. The clubs of Britain, which were once
the centre-points of English literature, and at which
there was indeed "feast of reason and flow of soul," sent
many a genius to an untimely grave, and even such men
as Johnson and Addison became charmed with the
pleasures of the table. The passions of Burns, Byron,
and many others, are too sorrowfully known to lead us
to multiply less noted examples. In noticing the biog-
raphies of great men, we have been too often surprised
to find how many of them were addicted to excesses in
eating and drinking, whose fair literary fame has over-
shadowed the memory of their evil habits, and whose
error is only slightly noticed in the more minute record
of their lives. "Died of over-feeding," would be a
truthful epitaph on the stone of many a one who has the
credit, falsely so called, of dying from a "too earnest devotion to literary pursuits."

On the other hand, Aristotle boasted that he could
have lived when acorns were the food of men. Newton
seemed to eat only that he might sustain life, and Shelly
and Milton belonged to the frugal class. Experience
shows that Swift was right when he declared "temperance to be a necessary virtue for great men," and
enough of good food and good exercise are no unimportant elements of greatness.

Letter III.

To Mechanics, Artizans and Laborers generally.

So far as you may be addressed in a body, it may
safely be said that none are so likely to enjoy a good de-

gree of health. Your occupations require physical labor and exertion, which is of itself a source of good
health, and in many employments a guarantee of good
appetite, good digestion and contented life. Those who
are called upon to work in the open air, are also secured
the blessings of perfect ventilation, and, in fact, all the
indispensables of physical comfort are compatible with
your sphere in life. As a rule, none suffer so much as
this class from the want of proper ablution and bathing.
With their time sedulously employed, when not engaged they are busy resting, and are thus apt to avoid
even little exertions. Many of them who regard themselves as neat and cleanly, bathe the whole body but
seldom, and their clothing is worn until much soiled.
For two reasons none stand in so imperious need of frequent washings, as these. Their occupations usually
provide much dirt or dust settling about and upon them,
and as perspiration is frequently caused by their labor,
the sweat, conveying out the impure secretions, is dried
upon the surface of the skin or the clothing adjacent.
Beside the refreshment it imparts, bathing thus becomes
to them an important duty. The body may be sponged
beside a warm fire if the weather is cold, and that without all the modern improvements, and no greater luxury has the working man, than the comfort which suc-
ceeds. To you, especially, a thorough wash each week
will be found most conducive to health, and without it
you cannot do justice to the framework on whose power
you directly depend for sustenance. Then there are
special indications to be attended to in each particular
trade.

Those exposed to the sun, or high degrees or low degrees, or sudden changes of temperature, need to inform themselves as to the best methods of avoiding ill results. Masons and others who work in a constrained position, must guard against headaches, and congestive tendency to the brain; and thus every mechanic should ask himself the question, What are the diseases, or tendencies, or exposure, or influences of any kind detrimental to my health, to which my vocation subjects me? Attention thus drawn to the subject will enable you to avoid many of them, and guard against the effects of others when unavoidable; to draw the attention of superiors or head managers to such arrangements as will secure to those employed easy positions, fresh air, etc., and, where if need be, to call upon the physician, the chemist, or others cognizant of matters of sanitory interest, to furnish such direction as will prevent disease.

Many of the arts bring the laborer in contact with deleterious substances, and type-makers, washers of lead, paper-makers and manifold other artizans, are exposed from time to time to fumes, gases or confinement prejudicial to the welfare of the body. Now, very often, those who by their education are fitted to look into these matters, can afford such information or institute such investigations as will prove of material advantage both to employer and laborer. Sir Humphry Davy never dug a single scuttle of coal, but when thousands had been mangled and slain by the explosions of gas in the deep mines of England, he was employed to see if he could not divine some method by which these terrible calamities might be prevented. After the careful trial of mouths, he furnished the safety lamp, with which the miner could walk to and fro amid these caverns without fear of sudden catastrophe. We are satisfied that in many species of manufactory, physicians and others could suggest means and plans by which the condition of the workmen would be much improved. The use of a little sulphuric acid by workers in lead is a very simple matter, but often all greatly conducive to their health; and so, we doubt not, there are manifold other trades in which those employed might, by a careful observer and experimenter, be taught to overcome many sources of evil.

Every laborer is entitled to be placed in a position which is as conducive to his health as the peculiar occupation will permit, and the attention both of master and mechanic should be drawn to this. The value of labor as in itself productive capital, the economy of preventing disease among operatives, and the demands of natural benevolence, call for a more careful regard to this matter. Within the last few years, wonderful improvements have been made in many of the eastern manufactories, in order to secure health, and owners have found large expense amply repaid, not only in the increased comfort of those employed, but in the greater amount of labor executed. Monarchial government seems to have some advantage over a republic, when it sends its officers, its physicians and sanitary reformers into each workshop and factory, to see that its people have as far as may be, health secured, as the first reward of labor; but with us there is only need that the laborers recognize the want, and public opinion and the interests of employers, will soon secure the requisite arrangements. Let mechanics or artizans, to whatever class they belong, see to it that they do not unnecessarily sacrifice their physical well-being, when their own investigations or attentions, or the single visit of an intelligent physician to their workshop would furnish suggestions saving many an uncomfortable ailment, or prolonging the average life of tradesmen. Hundreds of pent-up factories, unhealthy localities chosen for their cheapness, undrained grounds, ill ventilated rooms, etc., all over our country, are sickening and shortening the lives of many who can illy be spared from their families, or from our industrial wealth, and I shall be happy if this brief letter should draw the attention of employer or artizan to the subject.

Letter IV.

To Public Officers of Towns, Cities, etc.

Those who occupy the positions of trust and authority, have many opportunities for promoting the general health of those under their jurisdiction. Beside the large class of contagious disorders, there are many others indirectly communicable by epidemics or by foul atmosphere, and hence the individual health of each member of community is concerned in that of the aggregate mass. There are many important and well-defined hygienic laws which have so direct a bearing upon the welfare of the people, that a knowledge of them is essential to good government, and many a one of them would fill a nobler place upon the statute-book than that occupied by points of minor importance. In our country, we think that there is a growing necessity for calling the attention of officials not only to legal sanitory enactments, but to the whole subject of sanitory reform; and even a superficial examination of the codes of public institutions, the standing armies, and the cities of Europe, exhibit a much greater perfection of discipline than prevails with us. The health-power of masses is with individuals, and hence the importance of a right understanding of nature's laws, by at least the heads of families; but where we cannot secure this "for the public good," it is right to require adherence to all the indispensable pre-requisites of public health. Public schools and institutions of every kind need the careful supervision of those fully alive to the value of correct hygienic management, codes of health, and directions as to what disposition shall be made of all useless material in private houses, might with advantage be distributed like tracts at the homes of all the people, and thus much good be accomplished. Those who exalt the physical condition only of the masses, deserve well to have their names enrolled upon the schedule of philanthropy, for the moral and the intellectual are so closely associated with bodily prosperity, that the welfare of mankind is thus greatly promoted. Epaminondas never showed more beautifully the excellency of his official character, than when serving as a city scavenger, and those in higher offices may well take him as a model for imitation. The rapid development and growth of our cities, sometimes upon land never cultivated, in no way properly drained, and surrounded by miasmatic influ-

ences, and besides the tendency which there is, in the
grasp after wealth, to overlook every thing which does
not induce present discomfort, often allows the accumulation of smaller evils until rendered intense by association, they show themselves in a general deterioration of
health or a sudden prevalence of alarming epidemic.
Disease is not always an accident, or even a direct visi-
tation of Providence; but oftener the direct effect of
determinate causes, a necessary result of a general disobedience of nature's laws, the high penalty we pay for
public neglects, and nothing is more a subject for de-
vout gratitude than the means which have been provided us to overcome and neutralize these evils. Winds
and waves, showers and frosts, freezing and fire, are appointed means of purification, the million leaves that
rustle in mid-air are catching up the noxious exhalations
of decomposition, and all nature is busy preserving an
equilibrium of health. Food, and coal, and lime, and
chlorine, and manifold provisions for vitality are manufactured in God's world-wide laboratory from the noxious things which might do us harm, and we have grand
encouragement to aid the processes of nature. Medical
science, at every step it takes into the back-ground of
causes, sees how much of public physical disaster results
from appreciable mistakes, and the correction of many
of these needs only the diffusion of intelligence and the
exercise of appointed authority. None have so great
power in this matter as the mayors and councils of cities,
and others whose business in ruling it is to endeavor to
diffuse the greatest good to the greatest number. The
most eminent of philanthropists first had his attention

drawn to this subject by a small official position, and there is yet opportunity for fame, and what is greater than fame, the inward satisfaction of elevating the condition of the race by the exercise of means within legitimate power for diffusing sanitory intelligence and applying sanitory law. May you who are in authority thus have the disposition to "magnify your office."

LETTER V.

To those in advancing years.

As life advances, the constitution undergoes certain important changes, and these, in order to secure a healthy old age, must be met by corresponding modifications in the modes of living. There are few who do not at this period suffer from mistakes in earlier life, or who have not formed habits which need to be restrained and controlled by reason and the teachings of experience. The degree of exercise needed at this period is very different from that in early life. The body does not so readily rally from fatigue, and hence exercise should be taken frequently rather than continuously, and should be such as not to produce exhaustion. A great error is made by many in advancing years, by substituting for a life of activity, one of indolent ease. Men naturally look forward to the time when they can live upon the gains of former days, when they may retire from business and enjoy themselves in partaking of the bounties of their early industry; but, alas! how unsatisfying is the realization of the hope. When they have gained their wonted desire, such soon learn that employment is necessary both for the health of body and mind, and often in new ventures seek for the occupation they crave. Many are the instances familiar to us, where a sudden change of this kind has been fraught with most serious consequences, and those thus studying their ease have often re-- tired into apoplexy, paralysis or mal-content. It is desirable, it is true, that at this age there should be, if cares and labors have been severe, some diminution, but not an entire cessation. It was but yesterday that I accosted an aged German farmer with the expression of surprise, that he should still continue to work so much. "Ah," said his son who stood by, "father has to work; for when he stops a few days he gets sick." There was philosophy in that reply. Sudden changes are unsafe; and the equlibrium to be sought is a medium between the extremes of forced exertion and indolent ease. Just enough to do, is the highest earthly jewel in the crown of human happiness. Labor which is mentally exciting, inactive exposure, the stooping posture and violent exertion, such as lifting, running, etc., are most to be avoided by those advancing in life. Frequent bathing, especially with the warm bath, is often at this period a very grateful exercise, and one which can be indulged when out-door exposure would be impracticable.

The aged need to guard especially against cold and sudden alterations of temperature. They are more susceptible to intense weather, and are easily chilled. Bronchitis, influenza, and the like, are less easily overcome by them than formerly, and hence they should only expose themselves when in motion. The blood is less active in its circulation, and the extremities, as the feet and hands, should be well protected by woolen covering. In general, the adult requires much less sleep than the younger. For this reason they should not retire too early, and rise soon after waking. Occupation for the mind is as important to the aged as for the body; not severe study, or anxiety, or continuous reading; but a mild interest in passing events, and a relish for conversation and society. Discouragement and misanthropy are sometimes disposed to steal their way amid the pains of declining life, and grief to hang heavily upon the heart. Nothing sooner brings gray hairs down to the grave, and hence whatever contributes to cheerfulness and happy contentment, should be sought.

The demand for food lessens with declining life, and many things easily digested in earlier and more active days, will be found to disagree. Less animal food is required, and milk and starchy varieties of food are often found most nutritious. Hearty suppers should especially be avoided, on account of tendencies to organic disease, and meat once a day will generally be found to suffice. I have myself known many instances in advanced life, where the threatenings of disease have led to an entire change as to the amount and quality of the things eaten, with a most happy effect. How commonly do diseases of the heart or of the brain prove suddenly fatal after a hearty meal, and hence none more than the aged need to be the masters of their appetites. There is something venerable and almost sublime in the human form bearing up buoyantly against the vicissitudes of time, with well sustained powers of body and

16

mind, and with the soul cheered with the hope, after a good old age, of a glorious immortality.

Rightly viewed, the duty of living as long as we can, is as imperative as any other; and the pleasures to be derived, and the good to be secured by a happy and healthy old age, are such as to form a strong argument for the proper regulation of appetites and passions in manhood and middle life, and for living wisely as we pass to its evening horizon.

> "So live that life shall be no burden;
> And when its setting sun throws shadows
> O'er our pathway, let them be such as
> Lend our days the evening landscape's charm,
> And tinge the heavens with radiance
> Emblematic of re-rising."

Letter VI.

To Physicians.

No class of men have better right to claim the privilege of exalting the conditions and circumstances of human life, than have the members of the medical profession. The whole study of our science introduces us at once to facts and principles which show the basis upon which must rest health and physical prosperity, while it cannot but expose to our view the grave errors which are too often the direct sources of aggravated and prevailing disease. Practice still more fully presents these individually to our view, and prompts to an examination of the causes of human suffering. The author of a "Life for a Life" says: "If I were inclined to quit the army, I believe the branch of my profession which I should take up would be that of sanitory reform—the study of health rather than of disease, of prevention rather than cure. It often seems to me, that we of the healing art commenced at the wrong end; that the energy we devote to the alleviation of irremedial diseases, would be better spent in the study and practice of means to preserve health." There is much force in this observation and it is all important for the reputation of our profession, that we assume a department of philanthropy, to the supervision of which we are justly entitled. Professions command the respect of mankind much in proportion as they exhibit a utilitarian, race-elevating element; and while none contribute more frequent gratuitous benefits to the sick and destitute, we need to take a step higher in the regions of social amelioration, and earnestly propagate and labor for those principles which will secure immunity from disease, as well as remedies for it. From the success which has attended the few who have labored in this department, we are satisfied that the principles of prevention are wider, nobler, more practical, and more determinate, even, than those of cure, and the apparent contradictions of medical men in their views as to laws of health, is due more to a neglect of investigation, than to any obscurity in the laws themselves. The new light thrown upon physiology and pathology in the last few years, is a foundation step in the right direction, but even yet the power of medical analysis has not been sufficiently brought to bear upon hygiene, diatetics, and sanitory principles in general. Correct views in the profession freely communicated to the public at large, on these and some other topics, cannot but ennoble ourselves, our practice, and our race. It has been well said, that "the true method of putting down pretenders, in medicine as in every thing else, is to enlighten the public mind. It is ignorance which affords patronage to secret remedies, miraculous cures and quackeries of all sorts, both in and out of the domain of physic. When all medical practitioners shall cease to be pretenders to more knowledge than they really possess, then will the public cease to patronize quackery; and a more complete education, intellectual, moral and professional, engendering higher views of their duties, will be productive of greater benefit and cause them to rank higher in the estimation of the public." That symmetry which God has stamped upon the universe, does not fail in the "human form divine," and causation is a law of the physical man, just as much as it is of other created things; and it is no more difficult to discover and develop principles and practice which will improve the aggregate of human health, than it is to elaborate and induce conformity to the laws of intellectual and moral progress. When amid the desolations of the plague a fire swept through London, and made such havoc with houses as the disease had with life, men stood appalled until they found that it was a heaven-sent purifier and the plague was stayed. It corrected the oversights of officers and of the medical staff, and taught a significant lesson about the connection, even of epidemics, with appreciable causes. From this and volumes since, of facts accumulated by the past and illustrated in the present, let us, as physicians, feel it our special duty and privilege to investigate the physical

laws of human society, and diffuse such information
over the world-wide dominion of suffering and death as
will ameliorate and elevate our common humanity
Such effort will be appreciated, the physician and his
profession will rise still higher in the scale of dignity
and esteem, as more than ever a living power, and more,
and better, and beyond all, the satisfaction of doing
good, the sweetest thing of life, will dwell within us as
an abiding comforter, fitting us for cheerful duty here,
and with divine enlightenment for happiness and peace
beyond.

Natural, Truthful, and Enticing

THE
HOMESTEAD ON THE HILLSIDE,
And Other Tales.

BY MRS. MARY J. HOLMES,

The Popular Author of "Tempest and Sunshine" and "The English Orphans."

In One Volume, 380 Pages, 12mo. Price $1 00.

The numerous and delighted readers of "TEMPEST AND SUNSHINE" and "THE EN-GLISH ORPHANS"—Mrs. Holmes' former works—will be pleased to learn that another work of their favorite author is again within their reach. That this work will be eagerly sought and widely read, her former brilliant success affords the surest guaranty. Mrs. Holmes is a peculiarly pleasant and fascinating writer. Her subjects are the home and family relations. She has the happy faculty of enlisting the sympathies and affections of her readers and of holding their attention to her pages, with deep and absorbing interest. **The Homestead on the Hillside** is, therefore, attracting the liveliest attention ; and readers and

REVIEWERS ARE DECIDED IN ITS PRAISE.

Any one taking up the book must take a "through ticket," as there is no stopping place "this side" of the last page. The arts of the designing woman are given in their true color, showing to what oily-tongued hypocrisy humanity will stoop for the furtherance of its purposes ; what a vast amount of unhappiness one individual may bring upon an otherwise happy family ; what untold misery may result from the groveling spirit of fancied revenge, when cherished in the bosom of its unhappy possessor.—*Brockport Gazette.*

The talented author of " Tempest and Sunshine" has again hit on a happy subject. "The Homestead on the Hillside" has afforded her ample scope for the exercise of those high descriptive powers and those striking portraitures of character which have rendered her former works such general favorites. In one word, the book before us is no ordinary production.—*Philadelphia Daily News.*

Vigor, variety, a boldness and freedom of style and expression, eccentricity alike of character and incident, are among its most striking peculiarities. She has improved, in the book before us, upon her first effort, and several of these tales will not fail to add to her already well established reputation as a vigorous and attractive writer.—*Bost. Atlas.*

The artfulness and resignation exhibited by the Widow Carter, in her modest but not unnatural endeavors to gain the tender regard of Mr. Hamilton, as she smoothed the pillow of his dying wife, deserve the especial attention of gentlemen liable to a like attempt from a similar cause. They will doubtless see a dozen widows in the very dress and position of the philanthropic Mrs. Carter. There is quite a moral for young Misses, too, in the book."—*N. Y. Dutchman.*

It cannot fail to please the lovers of flowing and graceful narrative.—*Tribune.*

It will be superfluous to say that Mrs. Holmes is a charming writer.—*True Flag.*

Its genial spirit, its ready wit, its kindly feeling, will doubtless meet with due appreciation from all its readers. It touches with ready sympathy the fountains of mirth and tears, and one can neither restrain the one nor withhold the other, in reading its tales of joy and sorrow.—*Broome Repub.*

We have perused this book with none but feelings of pleasure ; and we have closed its pages, bearing in our heart its sweet spirit and eloquent moral. We heartily commend it.—*Lockport Courier.*

Her portrayal of human character and actions are admirable ; her style is fluent and fascinating, and a most intense degree of interest is kept up throughout the volume. But among all its excellent qualities, most prominent appears its eloquent morals. Read it, so that you can have it to say, " I ONCE READ A GOOD BOOK."—*Lockport Democrat*

Sold by all Booksellers. Single copies sent by mail, *post paid*, upon receipt of the price.

C. M. SAXTON, BARKER & CO., Publishers,

25 *Park Row, New York.*

A Book which will not be forgotten.

'LENA RIVERS.

BY MARY J. HOLMES,

Author of "Tempest and Sunshine," "The English Orphans," "The Homestead on the Hillside," etc. etc.

In One Volume, 416 Pages, 12mo. Price $1 00.

As the social and domestic relations are the great sources of happiness, or its opposites, so those romances that properly treat of those relations—of the virtues that adorn, and of the vices that deform them —are clearly the most interesting, impressive, and useful.

'LENA RIVERS is an American Domestic Story, unveiling in a masterly manner the sources of social and domestic enjoyment, or of disquiet and misery. By intermarriages of New England and Kentucky parties, a field is opened to exhibit both *Yankee and Southern domestic life*, for which the talented authoress was well prepared, being of Yankee birth and early education, and having subsequently resided in the South She was thus especially fitted to daguerreotype the *strictly domestic and social* peculiarities of both sections.

'LENA RIVERS AND THE PRESS.

A work of unusual promise. Mrs. Holmes possesses an enviable talent in the study of American character, which is so perfectly developed by acute observation from life, that it would now be impossible for her to write an uninteresting book.—*Phila. Sat. Bulletin.*

There still lingers the artist-mind, enlivening, cheering, and consoling by happy thoughts and pleasant words; moving the heart alternately to joy or sorrow, convulsing with laughter, or bringing tears to the eyes.—*Rochester American.*

The characters are well drawn, and the tale is one of interest. It will find many well pleased readers.—*Albany Statesman.*

The story is simple, natural, truthful.—*Rochester Daily Advertiser.*

Before we were aware, we had read the first two chapters. We read on—and on—and it was long after midnight when we finished the volume. We could not leave it. We know of no work with which we could compare "'Lena Rivers"—so as to form a just estimation of its merits.—*Merrickville Chronicle.*

It is not the first of the author's works, but it is the best.—*State Register.*

To the sex we commend it, on the assurance of its merit, volunteered to us by ladies in whose critical acumen we have the fullest confidence.—*Buffalo Express.*

The story opens in New England, and is continued in Kentucky, with very lively and characteristic sketches of scenery and character in both States. It is both GOOD and INTERESTING.—*New York Daily Times.*

The moral of the plot is excellent. Cowardly virtue, as exhibited by 'Lena's father, may here learn a lesson without suffering his bitter experience; while the rashness of youth may be warned against desperate acts, before a perfect understanding is had.—*New Bedford Express.*

This is an American novel possessing merit far superior to many which have been published during the last two years. The delineations of character are neatly and accurately drawn, and the tale is a deeply interesting one, containing many and varied incidents, illustrative of the workings of the human mind, and of social and domestic life in different parts of this country. The lesson to be deduced from its pages is a profitable one—which is more than can be said of many novels of the day.—*Portfolio.*

The scene of this tale is in Kentucky, although New England figures in it somewhat, and New Englanders still more largely. It is written in a lively style, and the interest 's not allowed to flag till the story terminates. One of the best things in the book is its sly and admirable hits at American aristocracy. It quietly shows some of "the plebeian 'ocation," which have, early or late, been connected with the "first families," and gives us a peep behind the curtain into the private life of those who are often objects of envy.

Sold by all Booksellers. Single copies mailed, *post paid*, on receipt of the price. **C. M. SAXTON, BARKER & CO., Publishers,**

25 Park Row, New York.

LOUIS NAPOLEON,

AND THE

BONAPARTE FAMILY.

COMPRISING A

HISTORY OF THE FRENCH REVOLUTION,

THE CAREER OF NAPOLEON, THE RESTORATION OF THE BOUR-
BONS, THE REIGN OF LOUIS PHILLIPPE, THE LIFE AND CA-
REER OF LOUIS NAPOLEON, AND THE CAUSES, EVENTS,
AND CONSEQUENCES OF THE CRIMEAN WAR.

BY HENRY W. DE PUY,

AUTHOR OF " KOSSUTH AND HIS GENERALS," " ETHAN ALLEN," ETC.

One Volume, 457 pp. 12mo., with Steel Portraits of Louis
Napoleon and the Empress Eugenie.   Price $1 25.

The foregoing is an interesting and a reliable history of the Bonaparte family, from the dawn of its celebrity to the present time.   It
contains a biography, not only of Napoleon I., Napoleon III., and of the
other members and branches of that distinguished family, but also of
other prominent actors in French affairs, with such a sketch of French
history as is necessary to the proper connection and clear understanding of the work.

EXTRACTS FROM REVIEWERS.

The Bonaparte family is one of the most remarkable that has ever appeared on the
earth.   Its origin was so humble, its elevation so rapid and dazzling, its power so great,
its fall so signal and low, its re-appearance in the person of Louis Napoleon so unex-
pected and potent, and its future so portentous, that it at once arrests the attention of
the modern historian, and audaciously takes its place in the very foreground of his
canvas.

We are not aware that any author has before attempted to present the entire Bonaparte family in one concise, yet clear and satisfactory volume.   It is a work long needed,
and for which every intelligent person constantly feels a pressing necessity.   Hence we
heartily welcome the work before us.   Its method is excellent, its breadth and grasp
very remarkable, and the style lucid and brilliant.   The engravings are superior, and
type, paper, and binding excellent.—*Taunton Democrat.*

An interesting and instructive volume.   The author has given a graphic description
of the career of the great Napoleon, free from that excessive flattery which distisguishes
the work of Abbott ; and the scarcely less brilliant career of Louis Napoleon is set forth
with admirable succinctness and truthfulness.   The work comprises the history of
France, and in fact of Europe, from the revolution of '89 to the present time, of which
the misfortunes and successes of Louis Philippe form a most interesting chapter.   The
biographical notices of the most distinguished characters that participated in public affairs during that period, is also a valuable feature of the work.—*Dem. Expounder.*

The style of the author is popular and attractive, and his book blends the interest of
history with that of biography.   Portraits of the present Emperor and of the Empress
EUGENIE, finely engraved, adorn the volume, which is handsomely issued in all respects.
—*Boston Telegraph.*

The notices of the various members of the Bonaparte family are written with clear-
ness, as are also the sketches of Louis XVIII., Charles X., Louis Philippe, Theirs, La-
martine, Guizot, Abdel-Kader, and numerous others whose names are familiar with
French movements during the present century.   The outline of the Russian War is
impartially given, a commendation which may be generally accorded to the entire volume.—THOMAS FRANCIS MEAGHER.

Sold by all Booksellers.   Mailed, *post-paid*, to any address, upon receipt of price.

C. M. SAXTON, BARKER & CO., Publishers,

25 *Park Row, New York.*

www.ingramcontent.com/pod-product-compliance
Lightning Source LLC
Chambersburg PA
CBHW030909270326
41929CB00008B/629